General Physics I

Mehdi Rahmani-Andebili

General Physics I

Practice Problems, Methods, and Solutions

 Springer

Mehdi Rahmani-Andebili
Electrical and Computer Engineering
The University of Alabama
Tuscaloosa, AL, USA

ISBN 978-3-031-92861-1   ISBN 978-3-031-92862-8   (eBook)
https://doi.org/10.1007/978-3-031-92862-8

This Springer imprint is published by the registered company Springer Nature Switzerland AG
The registered company address is: Gewerbestrasse 11, 6330 Cham, Switzerland

Preface

General Physics I and II are two mandatory courses for all science and engineering majors and even for some non-engineering majors that are taught at universities and colleges worldwide for freshman students. This textbook, as the *General Physics I*, has been prepared for instructors as well as for students taking General Physics courses. In each chapter of the textbook, different types of problems and exercises have been presented that are categorized as follows:

- *Problems with detailed solution*: They have been designed to teach students the subjects in detail. Moreover, they have been categorized in different levels based on their difficulty levels (easy, normal, and hard) and calculation amounts (small, normal, and large). This classification helps students study the book in the most efficient way.
- *Partially solved exercises*: They have been designed to encourage students to practice problems while guiding them through the problem-solving procedure and hinting the required formulas.
- *Exercises with final answer*: They have been designed to encourage students to practice more by themselves while hinting them by the final answer as well as to help instructors to give tests or quizzes.

In the following, the description of each chapter of the textbook is briefly presented.

Chapters 1 and 2 cover the subjects concerned with vectors and coordinate systems including inner product (dot product), vector product (cross product), and heuristic approaches.

Chapters 3 and 4 teach linear kinematics including average velocity, relative motion, free fall and uniformly accelerated motion, and projectile motion.

Chapters 5 and 6 study linear dynamics including Newton's laws, Newton's laws on inclined surface, Newton's Laws in an elevator, centripetal force, momentum, position-dependent force, potential energy, conservative force, work-energy theorem, conservation of mechanical energy principle, and spring force and Hooke's law.

Chapters 7 and 8 review collision, center of mass, and theorem of Pappus.

Chapters 9 and 10 investigate rotational kinematics and rotational dynamics including moment of inertia, parallel axis theorem, rotational kinetic energy, conservation of mechanical energy principle, and work-energy theorem.

Chapters 11 and 12 are concerned with simple harmonic motion including equations of motion, velocity, acceleration, and force of a simple harmonic oscillator; kinetic, potential, and mechanical energy of a simple harmonic oscillator; and period and frequency of a simple harmonic motion.

Moreover, the subjects of the General Physics II are as follows:

- *Electrostatic*
- *Electrical Capacitance, Resistance, and Current*
- *Magnetic Field*
- *Electromagnetic Induction*
- *Thermodynamics and Fluids*
- *Transverse and Longitudinal Waves*
- *Light, Mirrors, and Lenses*

Since the textbook includes the basic and advanced problems with very detailed problem solutions, it can be used as a practicing study guide by students and as a supplementary teaching source by instructors. Moreover, since the problems and exercises have very detailed solutions, the textbook is helpful for under-prepared students. In addition, it is beneficial for knowledgeable students because it includes advanced problems and exercises.

In preparing the problems and solutions, care has been taken to use methods typically found in the primary instructor-recommended textbooks. By considering this key point, the textbook is in the direction of instructors' lectures, and the instructors will not see any untaught and unusual problem solutions in their students' answer sheets.

Tuscaloosa, AL   Mehdi Rahmani-Andebili

The Other Works Published by the Author

AC Electric Machines: Practice Problems, Methods, and Solutions

AC Electrical Circuit Analysis: Practice Problems, Methods, and Solutions, Second Edition

Advanced Electrical Circuit Analysis: Practice Problems, Methods, and Solutions, Second Edition

Applications of Artificial Intelligence in Planning and Operation of Smart Grid

Applications of Fuzzy Logic in Planning and Operation of Smart Grids

Calculus I: Practice Problems, Methods, and Solutions, Second Edition

Calculus II: Practice Problems, Methods, and Solutions

Calculus III: Practice Problems, Methods, and Solutions

DC Electric Machines, Electromechanical Energy Conversion Principles, and Magnetic Circuit Analysis: Practice Problems, Methods, and Solutions

DC Electrical Circuit Analysis: Practice Problems, Methods, and Solutions, Second Edition

Design, Control, and Operation of Microgrids in Smart Grids

Differential Equations: Practice Problems, Methods, and Solutions

Electromagnetics: Practice Problems, Methods, and Solutions

Feedback Control Systems Analysis and Design: Practice Problems, Methods, and Solutions

General Physics II: Practice Problems, Methods, and Solutions

Mathematics of Engineering and Science: Practice Problems, Methods, and Solutions

MATLAB Lessons, Examples, and Exercises: A Tutorial for Beginners and Experts

Planning and Operation of Electric Vehicles in Smart Grid

Planning and Operation of Plug-in Electric Vehicles: Technical, Geographical, and Social Aspects

Power System Analysis: Comprehensive Lessons

Power System Analysis: Practice Problems, Methods, and Solutions

Precalculus: Practice Problems, Methods, and Solutions, Second Edition

Operation of Smart Homes

Contents

1   Vectors and Coordinate Systems: Part A 1
   1.1   Inner Product (Dot Product) 1
   1.2   Vector Product (Cross Product) 3
   1.3   Heuristic Approaches 5
   References 6

2   Vectors and Coordinate Systems: Part B 7
   2.1   Inner Product (Dot Product) 7
   2.2   Vector Product (Cross Product) 9
   2.3   Heuristic Approaches 10
   References 12

3   Linear Kinematics: Part A 13
   3.1   Average Velocity 13
   3.2   Relative Motion 16
   3.3   Free Fall and Uniformly Accelerated Motion 17
   3.4   Projectile Motion 17
   3.5   Power 23
   References 23

4   Linear Kinematics: Part B 25
   4.1   Average Velocity 25
   4.2   Relative Motion 29
   4.3   Free Fall and Uniformly Accelerated Motion 30
   4.4   Projectile Motion 30
   4.5   Power 36
   References 37

5   Linear Dynamics: Part A 39
   5.1   Newton's Laws 39
   5.2   Newton's Laws on Inclined Surface 43
   5.3   Newton's Laws in an Elevator 46
   5.4   Centripetal Force 47
   5.5   Momentum 48
   5.6   Position-Dependent Force and Potential Energy 49
   5.7   Conservative Force 49
   5.8   Work-Energy Theorem 50
   5.9   Conservation of Mechanical Energy Principle 50
   5.10   Spring Force and Hooke's Law 51
   References 52

6   Linear Dynamics: Part B 53
    6.1   Newton's Laws 53
    6.2   Newton's Laws on Inclined Surface 60
    6.3   Newton's Laws in an Elevator 64
    6.4   Centripetal Force 65
    6.5   Momentum 66
    6.6   Position-Dependent Force and Potential Energy 67
    6.7   Conservative Force 69
    6.8   Work-Energy Theorem 69
    6.9   Conservation of Mechanical Energy Principle 70
    6.10   Spring Force and Hooke's Law 71
    References 72

7   Collision and Centre of Mass: Part A 73
    7.1   Collision 73
    7.2   Centre of Mass and Theorem of Pappus 76
    References 79

8   Collision and Centre of Mass: Part B 81
    8.1   Collision 81
    8.2   Centre of Mass and Theorem of Pappus 87
    References 91

9   Rotational Kinematics and Dynamics: Part A 93
    9.1   Rotational Kinematics 93
    9.2   Moment of Inertia 94
    9.3   Parallel Axis Theorem 97
    9.4   Rotational Dynamics 98
    9.5   Rotational Kinetic Energy 99
    9.6   Conservation of Mechanical Energy Principle 100
    9.7   Work-Energy Theorem 102
    References 103

10   Rotational Kinematics and Dynamics: Part B 105
    10.1   Rotational Kinematics 105
    10.2   Moment of Inertia 106
    10.3   Parallel Axis Theorem 108
    10.4   Rotational Dynamics 109
    10.5   Rotational Kinetic Energy 111
    10.6   Conservation of Mechanical Energy Principle 113
    10.7   Work-Energy Theorem 116
    References 118

11   Simple Harmonic Motion: Part A 119
    11.1   Equations of Motion, Velocity, Acceleration, and Force of a Simple
           Harmonic Oscillator 119
    11.2   Kinetic, Potential, and Mechanical Energy of a Simple Harmonic
           Oscillator 123
    11.3   Period and Frequency of a Simple Harmonic Oscillator 124
    References 126

12   Simple Harmonic Motion: Part B   127
     12.1   Equations of Motion, Velocity, Acceleration, and Force of a Simple
            Harmonic Oscillator   127
     12.2   Kinetic, Potential, and Mechanical Energy of a Simple Harmonic
            Oscillator   133
     12.3   Period and Frequency of a Simple Harmonic Oscillator   135
     References   137

Index   139

About the Author

Mehdi Rahmani-Andebili is an Assistant Professor with the Department of Electrical and Computer Engineering at the University of Alabama. He received his first M.Sc. and Ph.D. degrees in Electrical Engineering from Tarbiat Modares University and Clemson University in 2011 and 2016, respectively, and his second Ph.D. (2nd M.Sc.) degree in Physics and Astronomy from the University of Alabama in Huntsville in 2019. Moreover, he was a Postdoctoral Fellow at the Sharif University of Technology during 2016–2017. As a professor, he has taught many courses and labs, including Power System Analysis, Electric Machines, Feedback Control Systems Analysis and Design, Signals and Systems, Electromagnetics, Electronics, Digital Logic, Industrial Electronics, Renewable Distributed Generation and Storage, AC Electrical Circuits Analysis, DC Electrical Circuits Analysis, Electrical Circuits and Devices, Fundamental of Electrical Engineering, Essentials of Electrical Engineering Technology, and Algebra and Calculus-Based Physics. Dr. Rahmani-Andebili has over 300 single-author or first-author publications including journal papers, conference papers, textbooks, books, and book chapters. He is an IEEE Senior Member and a permanent reviewer of many reputable journals. His research areas include Smart Grid, Applications of Artificial Intelligence in Planning and Operation of Power Systems, Integration of Renewables and Energy Storages into Power Systems, Energy Scheduling and Demand-Side Management, Electric Vehicles and Distributed Generation, and Advanced Optimization Techniques in Power System Studies.

Abstract

In this chapter, the basic and advanced problems of vectors and coordinate systems are studied. The subjects include inner product (dot product), vector product (cross product), and heuristic approaches. Herein, different types of problems and exercises are presented that are categorized as follows.

- **Problems with detailed solution**: They have been designed to teach students the subjects in detail. Moreover, they have been categorized in different levels based on their difficulty levels (easy, normal, and hard) and calculation amounts (small, normal, and large).
- **Partially solved exercises**: They have been designed to encourage students to practice problems while guiding them through the problem-solving procedure and hinting the required formulas.
- **Exercises with final answer**: They have been designed to encourage students to practice more by themselves while hinting them by the final answer as well as to help instructors to give tests or quizzes.

1.1 Inner Product (Dot Product)

Problem

1.1. Calculate the angle between the two vectors below [1–4].

$$\vec{a} = 3\widehat{i} + 3\widehat{j} - 3\widehat{k}$$

$$\vec{b} = -2\widehat{i} + 2\widehat{j} - 2\widehat{k}$$

Difficulty level   ● Easy   ○ Normal   ○ Hard
Calculation amount   ● Small   ○ Normal   ○ Large

1) $\theta = \arccos \dfrac{1}{3}$

2) $\theta = \arccos \dfrac{1}{2}$

3) $\theta = \arccos \dfrac{1}{5}$

4) $\theta = \arccos \dfrac{1}{4}$

Partially Solved Exercise

Calculate the angle between the following two vectors.

$$\vec{a} = \widehat{i} + \widehat{j} + \widehat{k}$$

$$\vec{b} = 2\widehat{j} - \widehat{k}$$

Solution

The problem can be solved as follows.

$$\cos\theta = \frac{\vec{a}\cdot\vec{b}}{|\vec{a}||\vec{b}|} = \frac{(\quad\quad\quad\quad\quad)}{\left(\sqrt{\quad\quad}\right)\left(\sqrt{\quad\quad}\right)}$$

$$\Rightarrow \cos\theta = \frac{1}{(\quad)\times(\quad)} = (\quad)$$

$$\Rightarrow \theta = \arccos\frac{1}{\sqrt{15}}$$

Notes

In this problem, the relations below have been used.

The angle between the two vectors can be calculated as follows.

$$\cos\theta = \frac{\vec{a}\cdot\vec{b}}{|\vec{a}||\vec{b}|}$$

The inner product (dot product) of two vectors can be calculated as follows.

$$\vec{a}\cdot\vec{b} = \left(a_1\widehat{i} + a_2\widehat{j} + a_3\widehat{k}\right)\cdot\left(b_1\widehat{i} + b_2\widehat{j} + b_3\widehat{k}\right) = a_1b_1 + a_2b_2 + a_3b_3$$

The magnitude of a vector can be calculated as follows.

$$|\vec{a}| = \left|a_1\widehat{i} + a_2\widehat{j} + a_3\widehat{k}\right| = \sqrt{(a_1)^2 + (a_2)^2 + (a_3)^2}$$

Exercise

Calculate the angle between the following two vectors.

$$\vec{a} = 3\widehat{i} + 3\widehat{j} - 6\widehat{k}$$

$$\vec{b} = -2\widehat{i} - 2\widehat{j} - 2\widehat{k}$$

Final Answer

$\theta = 90°$

Problem

1.2. Determine the values of the parameter "q" so that the vector $\vec{A}$ is perpendicular to the vector $\vec{B}$.

$$\vec{A} = q\hat{i} + 3\hat{j} + \hat{k}$$

$$\vec{B} = q\hat{i} - q\hat{j} + 2\hat{k}$$

Difficulty level 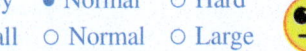   ○ Easy   ● Normal   ○ Hard
Calculation amount   ● Small   ○ Normal   ○ Large

1) 1, 0
2) 2, 1
3) 3, 1
4) 3, 2

Exercise

Calculate the values of the parameter "a" so that the vector $\vec{A}$ is perpendicular to the vector $\vec{B}$.

$$\vec{A} = a\hat{i} + \hat{j} - 2\hat{k}$$

$$\vec{B} = a\hat{i} - 2\hat{j} + \hat{k}$$

Final Answer

$a = -2, 2$

1.2   Vector Product (Cross Product)

Problem

1.3. Determine the vector product (cross product) of the following vectors.

$$\vec{a} = 2\hat{i} - 4\hat{j} + \hat{k}$$

$$\vec{b} = 6\hat{i} + 8\hat{j} - 15\hat{k}$$

Difficulty level    ● Easy   ○ Normal   ○ Hard
Calculation amount   ● Small   ○ Normal   ○ Large

1) $52\hat{i} + 36\hat{j} + 40\hat{k}$
2) $5\hat{i} + 36\hat{j} + 40\hat{k}$
3) $52\hat{i} + 6\hat{j} + 40\hat{k}$
4) $52\hat{i} + 36\hat{j} + 4\hat{k}$

Partially Solved Exercise

Calculate the vector product of the vectors below.

$$\vec{a} = \hat{i} - \hat{j} + \hat{k}$$

$$\vec{b} = \hat{i} + \hat{j} + \hat{k}$$

Solution

The problem can be solved as follows.

$$\vec{a} \times \vec{b} = \begin{vmatrix} \hat{i} & \hat{j} & \hat{k} \\ 1 & -1 & 1 \\ 1 & 1 & 1 \end{vmatrix} = (\qquad)\hat{i} + (\qquad)\hat{j} + (\qquad)\hat{k}$$

$$\Rightarrow \vec{a} \times \vec{b} = (\quad)\hat{i} + (\quad)\hat{j} + (\quad)\hat{k}$$

Notes

In this problem, the relation below has been used.

The vector product (cross product) of two vectors can be calculated as follows.

$$\vec{a} \times \vec{b} = \begin{vmatrix} \hat{i} & \hat{j} & \hat{k} \\ a_1 & a_2 & a_3 \\ b_1 & b_2 & b_3 \end{vmatrix} = (a_2 b_3 - a_3 b_2)\hat{i} + (a_3 b_1 - a_1 b_3)\hat{j} + (a_1 b_2 - a_2 b_1)\hat{k}$$

Exercise

Calculate the vector product of the vectors below.

$$\vec{a} = \hat{i} + \hat{j} + \hat{k}$$

$$\vec{b} = 2\hat{i} + 2\hat{j} + 2\hat{k}$$

Final Answer

$$\vec{a} \times \vec{b} = 0$$

Problem

1.4. Determine a vector that is perpendicular to the common plain of the vectors below.

$$\vec{A} = 2\hat{i} - 3\hat{j} + \hat{k}$$

$$\vec{B} = -3\hat{j} + 2\hat{k}$$

Difficulty level   ● Easy   ○ Normal   ○ Hard
Calculation amount   ● Small   ○ Normal   ○ Large

1) $-3\widehat{i} + 4\widehat{j} + 6\widehat{k}$

2) $2\widehat{i} + 3\widehat{j} + 5\widehat{k}$

3) $3\widehat{i} + 4\widehat{j} + 6\widehat{k}$

4) $-3\left(\widehat{i} - 4\widehat{j} + 2\widehat{k}\right)$

Exercise

Determine a vector that is perpendicular to the common plain of the following vectors.

$$\vec{A} = 2\widehat{i} - \widehat{j} + \widehat{k}$$

$$\vec{B} = \widehat{i} + \widehat{k}$$

Final Answer

$$\vec{N} = -\widehat{i} - \widehat{j} + \widehat{k}$$

1.3   Heuristic Approaches

Problem

1.5. Which one of the following choices is related to an object that its distance from the origin is being decreased.

Difficulty level   ○ Easy   ○ Normal   ● Hard

Calculation amount   ● Small   ○ Normal   ○ Large

1) $xv_y - yv_x < 0$

2) $v_x < 0, v_y < 0$

3) $xv_y + yv_x < 0$

4) $xv_x + yv_y < 0$

Problem

1.6. An object is moving on the $x - y$ plain where its parametric equation of motion is as follows.

$$\begin{cases} x(t) = R(\omega t - \sin \omega t) \\ y(t) = R(1 - \cos \omega t) \end{cases}$$

Determine the angle between its velocity and acceleration vectors at $t = 4s$.

Difficulty level   ○ Easy   ○ Normal   ● Hard

Calculation amount   ○ Small   ○ Normal   ● Large

1) $\frac{\omega}{3}$

2) $\frac{\omega}{2}$

3) ω

4) 2ω

References

1. Rahmani-Andebili, M. (2023). Calculus III – Practice Problems, Methods, and Solutions, Springer Nature.
2. Rahmani-Andebili, M. (2023). Calculus II – Practice Problems, Methods, and Solutions, Springer Nature.
3. Rahmani-Andebili, M. (2023). Calculus I (2nd Ed.) – Practice Problems, Methods, and Solutions, Springer Nature.
4. Rahmani-Andebili, M. (2024). Precalculus (2nd Ed.) – Practice Problems, Methods, and Solutions, Springer Nature.

Vectors and Coordinate Systems: Part B

Abstract

In this chapter, the problems of the first chapter are fully solved, in detail, step-by-step, and with different methods.

2.1 Inner Product (Dot Product)

2.1. Based on the information given in the problem, we have:

$$\vec{a} = 3\widehat{i} + 3\widehat{j} - 3\widehat{k}$$

$$\vec{b} = -2\widehat{i} + 2\widehat{j} - 2\widehat{k}$$

The problem can be solved as follows [1–4].

$$\cos\theta = \frac{\vec{a}.\vec{b}}{|\vec{a}||\vec{b}|} = \frac{3 \times (-2) + 3 \times 2 + (-3) \times (-2)}{\left(\sqrt{3^2 + 3^2 + (-3)^2}\right)\left(\sqrt{(-2)^2 + 2^2 + (-2)^2}\right)}$$

$$\Rightarrow \cos\theta = \frac{6}{3\sqrt{3} \times 2\sqrt{3}} = \frac{1}{3}$$

$$\Rightarrow \theta = \arccos\frac{1}{3}$$

Choice (1) is the answer.

© The Author(s), under exclusive license to Springer Nature Switzerland AG 2025
M. Rahmani-Andebili, *General Physics I*, https://doi.org/10.1007/978-3-031-92862-8_2

Notes

In this problem, the relations below have been used.

The angle between the two vectors can be calculated as follows.

$$\cos\theta = \frac{\vec{a}\cdot\vec{b}}{\left|\vec{a}\right|\left|\vec{b}\right|}$$

The inner product (dot product) of two vectors can be calculated as follows.

$$\vec{a}\cdot\vec{b} = \left(a_1\widehat{i} + a_2\widehat{j} + a_3\widehat{k}\right)\cdot\left(b_1\widehat{i} + b_2\widehat{j} + b_3\widehat{k}\right) = a_1b_1 + a_2b_2 + a_3b_3$$

The magnitude of a vector can be calculated as follows.

$$\left|\vec{a}\right| = \left|a_1\widehat{i} + a_2\widehat{j} + a_3\widehat{k}\right| = \sqrt{\left(a_1\right)^2 + \left(a_2\right)^2 + \left(a_3\right)^2}$$

2.2. Based on the information given in the problem, we have:

$$\vec{A} = q\widehat{i} + 3\widehat{j} + \widehat{k}$$

$$\vec{B} = q\widehat{i} - q\widehat{j} + 2\widehat{k}$$

As we know, two vectors are perpendicular to each other if their inner product (dot product) is zero. In other words:

$$\vec{A}\cdot\vec{B} = 0$$

Therefore:

$$\left(q\widehat{i} + 3\widehat{j} + \widehat{k}\right)\cdot\left(q\widehat{i} - q\widehat{j} + 2\widehat{k}\right) = 0$$

$$\Rightarrow q^2 - 3q + 2 = 0 \Rightarrow (q-2)(q-1) = 0$$

$$\Rightarrow q = 1, 2$$

Choice (2) is the answer.

2.2   Vector Product (Cross Product)

2.3. Based on the information given in the problem, we have:

$$\vec{a} = 2\widehat{i} - 4\widehat{j} + \widehat{k}$$

$$\vec{b} = 6\widehat{i} + 8\widehat{j} - 15\widehat{k}$$

The problem can be solved as follows.

$$\vec{a} \times \vec{b} = \begin{vmatrix} \widehat{i} & \widehat{j} & \widehat{k} \\ 2 & -4 & 1 \\ 6 & 8 & -15 \end{vmatrix} = (60 - 8)\widehat{i} + (6 + 30)\widehat{j} + (16 + 24)\widehat{k}$$

$$\Rightarrow \vec{a} \times \vec{b} = 52\widehat{i} + 36\widehat{j} + 40\widehat{k}$$

Choice (1) is the answer.

> **Notes**
>
> In this problem, the relation below has been used.
>
> The vector product (cross product) of two vectors can be calculated as follows.
>
> $$\vec{a} \times \vec{b} = \begin{vmatrix} \widehat{i} & \widehat{j} & \widehat{k} \\ a_1 & a_2 & a_3 \\ b_1 & b_2 & b_3 \end{vmatrix} = (a_2 b_3 - a_3 b_2)\widehat{i} + (a_3 b_1 - a_1 b_3)\widehat{j} + (a_1 b_2 - a_2 b_1)\widehat{k}$$

■ ■ ■

2.4. Based on the information given in the problem, we have:

$$\vec{A} = 2\widehat{i} - 3\widehat{j} + \widehat{k}$$

$$\vec{B} = -3\widehat{j} + 2\widehat{k}$$

As we know, the vector resulted from the cross product of two vectors is perpendicular to the common plain of the two vectors (normal vector). Therefore:

$$\vec{N} = \vec{A} \times \vec{B} = \begin{vmatrix} \widehat{i} & \widehat{j} & \widehat{k} \\ 2 & -3 & 1 \\ 0 & -3 & 2 \end{vmatrix} = -3\widehat{i} - 4\widehat{j} - 6\widehat{k}$$

Moreover, a normal vector with any scalar factor is perpendicular to the common plain of the two vectors. Thus:

$$-\left(\vec{A} \times \vec{B}\right) = 3\hat{i} + 4\hat{j} + 6\hat{k}$$

Choice (3) is the answer.

Notes

In this problem, the relations below have been used.

The normal vector of a plane, made by two vectors, can be calculated from their cross product. In other words:

$$\vec{N} = \vec{A} \times \vec{B}$$

The vector product (cross product) of two vectors can be calculated as follows.

$$\vec{a} \times \vec{b} = \begin{vmatrix} \hat{i} & \hat{j} & \hat{k} \\ a_1 & a_2 & a_3 \\ b_1 & b_2 & b_3 \end{vmatrix} = (a_2 b_3 - a_3 b_2)\hat{i} + (a_3 b_1 - a_1 b_3)\hat{j} + (a_1 b_2 - a_2 b_1)\hat{k}$$

2.3   Heuristic Approaches

2.5. The distance of an object (with the position and velocity vectors $\vec{r}(t)$ and $\vec{v}(t)$, respectively) from the origin decreases if $\vec{r}(t) \cdot \vec{v}(t) < 0$. Therefore:

$$\vec{r}(t) \cdot \vec{v}(t) = \left(x(t)\,\vec{i} + y(t)\,\vec{j}\right) \cdot \left(v_x(t)\,\vec{i} + v_y(t)\,\vec{j}\right) = x(t)v_x(t) + y(t)v_y(t)$$

$$\Rightarrow xv_x + yv_y < 0$$

Choice (4) is the answer.

Notes

In this problem, the relation below has been used.

The inner product (dot product) of two vectors can be calculated as follows.

$$\vec{a} \cdot \vec{b} = \left(a_1\hat{i} + a_2\hat{j} + a_3\hat{k}\right) \cdot \left(b_1\hat{i} + b_2\hat{j} + b_3\hat{k}\right) = a_1 b_1 + a_2 b_2 + a_3 b_3$$

2.6. Based on the information given in the problem, we have:

$$\begin{cases} x(t) = R(\omega t - \sin \omega t) \\ y(t) = R(1 - \cos \omega t) \end{cases}$$

The problem can be solved as follows.

$$v_x = \frac{d}{dt}x = R\omega(1 - \cos \omega t)$$

$$a_x = \frac{d}{dt}v_x = R\omega^2 \sin \omega t$$

$$v_y = \frac{d}{dt}y = R\omega \sin \omega t$$

$$a_y = \frac{d}{dt}v_y = R\omega^2 \cos \omega t$$

The angle between the velocity vector and the positive direction of x-axis can be calculated as follows.

$$\tan \theta_v = \frac{v_y}{v_x} = \frac{R\omega \sin \omega t}{R\omega(1 - \cos \omega t)} = \frac{\sin \omega t}{1 - \cos \omega t}$$

$$\Rightarrow \tan \theta_v = \frac{2 \sin \frac{\omega t}{2} \cos \frac{\omega t}{2}}{2 \sin^2 \frac{\omega t}{2}} = \frac{\cos \frac{\omega t}{2}}{\sin \frac{\omega t}{2}} = \cot \frac{\omega t}{2} = \tan\left(\frac{\pi}{2} - \frac{\omega t}{2}\right)$$

$$\Rightarrow \theta_v = \left(\frac{\pi}{2} - \frac{\omega t}{2}\right)$$

Likewise, the angle between the acceleration vector and the positive direction of x-axis can be calculated as follows.

$$\tan \theta_a = \frac{a_y}{a_x} = \frac{R\omega^2 \cos \omega t}{R\omega^2 \sin \omega t} = \frac{\cos \omega t}{\sin \omega t}$$

$$\Rightarrow \tan \theta_a = \cot \omega t = \tan\left(\frac{\pi}{2} - \omega t\right)$$

$$\Rightarrow \theta_a = \left(\frac{\pi}{2} - \omega t\right)$$

Thus, the angle between the velocity and acceleration vectors is as follows.

$$\alpha = \theta_v - \theta_a = \left(\frac{\pi}{2} - \frac{\omega t}{2}\right) - \left(\frac{\pi}{2} - \omega t\right) = \frac{\omega t}{2}$$

At $t = 4s$, we have:

$$\alpha = 2\omega$$

Choice (4) is the answer.

Notes

In this problem, the relations below have been used.

The relation between the x component of the position and velocity of an object is as follows.

$$v_x = \frac{d}{dt}x$$

Likewise, the relation between the y component of the position and velocity of an object is as follows.

$$v_y = \frac{d}{dt}y$$

Moreover, the relation between the x component of the velocity and acceleration of an object is as follows.

$$a_x = \frac{d}{dt}v$$

Likewise, the relation between the y component of the velocity and acceleration of an object is as follows.

$$a_y = \frac{d}{dt}v$$

The angle between the vector $\vec{A}$ and the positive direction of x-axis can be calculated as follows.

$$\theta_A = \tan^{-1}\frac{A_y}{A_x}$$

Moreover:

$$\frac{d}{dt}\sin at = a\cos at$$

$$\frac{d}{dt}\cos at = -a\sin at$$

$$\sin 2t = 2\sin t\cos t$$

$$1 - \cos 2t = 2\sin^2 t$$

$$\cot t = \frac{\cos t}{\sin t}$$

$$\cot t = \tan\left(\frac{\pi}{2} - t\right)$$

References

1. Rahmani-Andebili, M. (2023). Calculus III – Practice Problems, Methods, and Solutions, Springer Nature.
2. Rahmani-Andebili, M. (2023). Calculus II – Practice Problems, Methods, and Solutions, Springer Nature.
3. Rahmani-Andebili, M. (2023). Calculus I (2nd Ed.) – Practice Problems, Methods, and Solutions, Springer Nature.
4. Rahmani-Andebili, M. (2024). Precalculus (2nd Ed.) – Practice Problems, Methods, and Solutions, Springer Nature.

Linear Kinematics: Part A

3

Abstract

In this chapter, the basic and advanced problems of linear kinematics are studied. The subjects include average velocity, relative motion, free fall and uniformly accelerated motion, trajectory equation of a projectile, projectile motion, and power. Herein, different types of problems and exercises are presented that are categorized as follows.

- **Problems with detailed solution**: They have been designed to teach students the subjects in detail. Moreover, they have been categorized in different levels based on their difficulty levels (easy, normal, and hard) and calculation amounts (small, normal, and large).
- **Partially solved exercises**: They have been designed to encourage students to practice problems while guiding them through the problem-solving procedure and hinting the required formulas.
- **Exercises with final answer**: They have been designed to encourage students to practice more by themselves while hinting them by the final answer as well as to help instructors to give tests or quizzes.

3.1 Average Velocity

Problem

3.1. Calculate the average velocity of an object after 30 s that its velocity-time curve is shown in Fig. 3.1 [1–4].

Difficulty level ● Easy ○ Normal ○ Hard
Calculation amount ● Small ○ Normal ○ Large

1) 10 *m/s*
2) 0 *m/s*
3) 15 *m/s*
4) 20 *m/s*

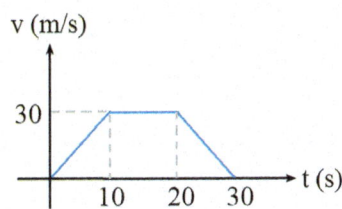

Fig. 3.1. The figure concerned with Problem 3.1

Exercise

Calculate the average velocity of an object after 10 s with the velocity-time curve shown in Fig. 3.2.

Final Answer

$\bar{v} = 2.5 \, m/s$

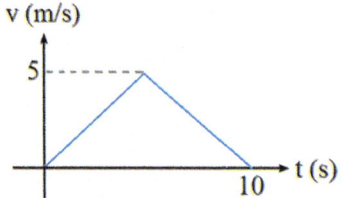

Fig. 3.2. The figure concerned with the exercise

Problem

3.2. The velocity-time curve of two objects is shown in Fig. 3.3. Which one of the following choices is correct about their average velocities during the interval $[t_1, t_2]$.

Difficulty level   ● Easy   ○ Normal   ○ Hard
Calculation amount   ● Small   ○ Normal   ○ Large
1) $\overline{v_A} = \overline{v_B}$
2) $\overline{v_A} > \overline{v_B}$
3) $\overline{v_A} < \overline{v_B}$
4) More data is needed to determine it.

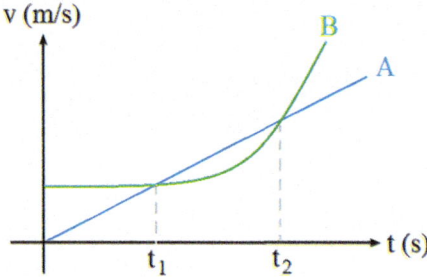

Fig. 3.3. The figure concerned with Problem 3.2

Problem

3.3. Calculate the average velocity of an object (m/s) that moves from $(x, y) = (2, 5)$ to $(5,9)$ for two seconds.

Difficulty level   ● Easy   ○ Normal   ○ Hard
Calculation amount   ● Small   ○ Normal   ○ Large
1) 5
2) 10
3) 2.5
4) 0

Partially Solved Exercise

Calculate the average velocity of an object that moves from $(x, y) = (0, 0)$ to $(3,4)$ during five seconds.

Solution

The problem can be solved as follows:

$$\bar{v} = \frac{\Delta r}{\Delta t} = \frac{\sqrt{(x_2 - x_1)^2 + (y_2 - y_1)^2}}{\Delta t}$$

$$\Rightarrow \bar{v} = \frac{\sqrt{(3 - 0)^2 + (4 - 0)^2}}{(\quad)} = \frac{(\quad)}{(\quad)}$$

$$\Rightarrow \bar{v} = 1 \, \frac{m}{s}$$

Problem

3.4. An object with velocity v, $2v$, and $3v$ moves for t, $2t$, and $3t$ time unit, respectively. Parametrically calculate the average velocity of the object.

Difficulty level   ○ Easy   ● Normal   ○ Hard
Calculation amount   ○ Small   ● Normal   ○ Large

1) $\frac{3}{7}v$

2) $\frac{9}{17}v$

3) $\frac{14}{9}v$

4) $\frac{7}{3}v$

Exercise

Calculate the average velocity of an object if it moves with the velocity of 2 m/s and 3 m/s for 15 s and 5 s, respectively.

Final Answer

$\bar{v} = 2.25 \, m/s$

Problem

3.5. An object with velocity v, $2v$, and $5v$ moves the distances x, $2x$, and $5x$, respectively. Parametrically calculate the average velocity of the object.

Difficulty level   ○ Easy   ● Normal   ○ Hard
Calculation amount   ○ Small   ● Normal   ○ Large

1) $2v$

2) $8v$

3) $\frac{8}{3}v$

4) $\frac{3}{8}v$

Partially Solved Exercise

An object moves 40% of the distance with a velocity of 20 *m/s* and the rest with 60 *m/s* velocity. Calculate the average velocity of the object (*m/s*).

Solution

The problem can be solved as follows:

$$\bar{v} = \frac{\Delta r_1 + \Delta r_2}{\dfrac{\Delta x_1}{v_1} + \dfrac{\Delta x_2}{v_2}}$$

$$\Rightarrow \bar{v} = \frac{(\quad) + (\quad)}{\dfrac{(\quad)}{(\quad)} + \dfrac{(\quad)}{(\quad)}}$$

$$\Rightarrow \bar{v} = \frac{(\quad)}{(\quad)}$$

$$\Rightarrow \bar{v} = \frac{100}{3} \ m/s$$

Problem

3.6. An object is moving with the velocity 20 *m/s* on a line with the equation $y = \sqrt{3}x + 1$. Determine the distance that it will move on the x-axis during three second.

Difficulty level   ○ Easy   ○ Normal   ● Hard
Calculation amount   ○ Small   ● Normal   ○ Large

1) 30 *m*
2) $30\sqrt{3}$ *m*
3) $60\sqrt{3}$ *m*
4) 60 *m*

3.2   Relative Motion

Problem

3.7. Two bullets with velocities of 10 *m/s* and 20 *m/s*, respectively, from the top and bottom of a 300 *m* height tower are simultaneously shot towards each other. Determine the time after $t = 0$ that they will meet one another.

Difficulty level   ○ Easy   ○ Normal   ● Hard
Calculation amount   ● Small   ○ Normal   ○ Large

1) 5 *s*
2) 10 *s*
3) 15 *s*
4) 20 *s*

Exercise

Two bullets with velocities of 10 m/s and 20 m/s are simultaneously shot upward from the bottom of a 300 m height tower. Calculate their distance after 0.5 s.

Final Answer

$d = 5\ m$

3.3   Free Fall and Uniformly Accelerated Motion

Problem

3.8. A person shoots a bullet with a velocity of 450 m/s towards a target at the distance of 45 m. What should be the target point so that the bullet hits the real target? Herein, assume that $g = 10\ m/s^2$.

Difficulty level   ○ Easy   ○ Normal   ● Hard
Calculation amount   ○ Small   ● Normal   ○ Large

1) 5 cm above the real target point.
2) 10 cm above the real target point.
3) 15 cm below the real target point.
4) 5 cm below the real target point.

3.4   Projectile Motion

Problem

3.9. A projectile is shot at an angle of 30° with respect to the ground. Then, the experiment is repeated at an angle of 60°. Calculate the ratio of ranges of the projectile in the second experiment to the first one.

Difficulty level   ● Easy   ○ Normal   ○ Hard
Calculation amount   ● Small   ○ Normal   ○ Large

1) $\sqrt{3}$
2) 1
3) $\dfrac{\sqrt{3}}{3}$
4) $\dfrac{1}{3}$

Problem

3.10. Consider Problem 3.9 and calculate the ratio of height of the peak points (apex) of the projectile in the second experiment to the first experiment.

Difficulty level   ● Easy   ○ Normal   ○ Hard
Calculation amount   ● Small   ○ Normal   ○ Large

1) 3
2) 1
3) $\dfrac{\sqrt{3}}{3}$
4) $\dfrac{1}{3}$

Partially Solved Exercise

The range of a projectile is twice the height of its peak point. Calculate the tangent of the projection angle.

Solution

The problem can be solved as follows.

$$R = 2h$$

$$\Rightarrow \frac{(\qquad)}{(\qquad)} = 2 \times \frac{(\qquad)}{(\qquad)}$$

$$\Rightarrow (\qquad) = \sin^2 \alpha$$

$$\Rightarrow (\qquad) = \sin^2 \alpha$$

$$\Rightarrow \tan \alpha = 2$$

Notes

In this problem, the relations below are needed.

$$\sin 2\alpha = 2 \sin \alpha \cos \alpha$$

$$\frac{\sin \alpha}{\cos \alpha} = \tan \alpha$$

Exercise

A projectile is shot at an angle of 30° with respect to the ground. Then, the experiment is repeated at an angle of 45°. Calculate the ratio of height of the peak points of the projectile in the second experiment to the first one.

Final Answer

$$\frac{h_2}{h_1} = 2$$

Exercise

What are the coordinates of the peak point of a projectile if its primary velocity and projection angle are v_0 and α, respectively?

Final Answer

$$\left(x_{peak}, y_{peak} \right) = \left(\frac{v_0^2 \sin 2\alpha}{2g}, \frac{v_0^2 \sin^2 \alpha}{2g} \right)$$

Exercise

What is the relation between the coordinates of the peak point of a projectile if its primary velocity and projection angle are v_0 and α, respectively?

Final Answer

$$y_{peak} = 0.5 \tan \alpha x_{peak}$$

Problem

3.11. What projection angle will result in the maximum height of the peak point of a projectile?

Difficulty level   ● Easy   ○ Normal   ○ Hard

Calculation amount   ● Small   ○ Normal   ○ Large

1) 0°
2) 30°
3) 45°
4) 90°

Problem

3.12. What projection angle will result in the maximum range of a projectile?

Difficulty level   ● Easy   ○ Normal   ○ Hard

Calculation amount   ● Small   ○ Normal   ○ Large

1) 0°
2) 30°
3) 45°
4) 90°

Exercise

Calculate the time that a projectile reaches its peak point if its primary velocity and projection angle are v_0 and α, respectively.

Final Answer

$$t_{peak} = \frac{v_0 \sin \alpha}{g}$$

Exercise

Calculate the time that a projectile reaches its range if its primary velocity and projection angle are v_0 and α, respectively.

Final Answer

$$t_{range} = \frac{2v_0 \sin \alpha}{g}$$

Problem

3.13. The primary velocity of a projectile is twice its velocity at the peak point. Calculate the projection angle of the projectile.

Difficulty level   ○ Easy   ○ Normal   ● Hard

Calculation amount   ● Small   ○ Normal   ○ Large

1) 30°
2) 60°
3) 45°
4) 15°

Problem

3.14. The primary velocity of a projectile and its velocity at the peak point are 30 *m/s* and 10 *m/s*, respectively. Calculate the height of the apex (*m*). Herein, assume that $g = 10 \ m/s^2$.

Difficulty level   ○ Easy   ○ Normal   ● Hard
Calculation amount   ● Small   ○ Normal   ○ Large
1) 45
2) 40
3) 20
4) 25

Exercise

The initial velocity of a projectile and its velocity at the peak point (apex) are 5 *m/s* and 4 *m/s*, respectively. Calculate the height of the apex (*m*). Herein, assume that $g = 10 \ m/s^2$.

Final Answer

$h = 0.45 \ m$

Exercise

Regarding a projectile motion, what is the relation between the initial velocity (v_0), the velocity at peak point (v_{peak}), the velocity before hitting the ground (v_{range}), and the minimum velocity (v_{min}) of a projectile? Herein, the air resistance is neglected.

Final Answer

$v_0 = v_{range} > v_{peak} = v_{min}$

Exercise

At which points, the angle between the acceleration and velocity of a projectile motion is maximum and minimum, respectively?

Final Answer

At the beginning and end of a projectile motion, respectively.

Exercise

At which point, the angle between the acceleration and velocity of a projectile motion is 90°?

Final Answer

At peak point.

Problem

3.15. The kinetic energy of a projectile at the peak point (apex) is 75% of its initial mechanical energy. Calculate the projection angle (radian).

Difficulty level   ○ Easy   ○ Normal   ● Hard
Calculation amount   ○ Small   ● Normal   ○ Large

1) $\dfrac{\pi}{6}$

2) $\dfrac{\pi}{3}$

3) $\dfrac{\pi}{4}$

4) $\dfrac{\pi}{5}$

Partially Solved Exercise

The kinetic energy of a projectile at the peak point (apex) is 25% of its primary energy. Calculate the projection angle (degree).

Solution

As we know, the velocity at peak point is equal to v_{0x} because $v_y = 0$ and $v_x = v_{0x} = $ Constant. Hence:

$$\frac{1}{2}mv_{0x}^2 = 0.25\left(\frac{1}{2}mv_0^2\right)$$

$$\Rightarrow v_{0x}^2 = (\qquad)v_0^2$$

$$\Rightarrow (v_0\cos\alpha)^2 = (\qquad)v_0^2$$

$$\Rightarrow \cos^2\alpha = (\qquad)$$

$$\Rightarrow \cos\alpha = (\qquad)$$

$$\Rightarrow \alpha = \frac{\pi}{3}\ rad$$

Notes

In this problem, the relations below have been used.

$$K = \frac{1}{2}mv^2$$

$$v_x = v\cos\theta$$

$$\cos\frac{\pi}{3} = \frac{1}{2}$$

Problem

3.16. The primary kinetic energy of a projectile with the projection angle of 30° is 60 *J*. Calculate its kinetic energy at the peak point in Joules.

Difficulty level   ○ Easy   ○ Normal   ● Hard

Calculation amount   ○ Small   ● Normal   ○ Large

1) 15

2) 35

3) 45

4) 55

Problem

3.17. A projectile is shot from the origin at an angle of 45° with respect to the ground. If this projectile passes from the point $(x, y) = (24, 14)$, calculate its primary velocity (*m/s*). Herein, assume that $g = 10 \ m/s^2$.

Difficulty level   ○ Easy   ○ Normal   ● Hard

Calculation amount   ○ Small   ● Normal   ○ Large

1) 10

2) 14

3) 24

4) 12

Partially Solved Exercise

A projectile is shot from the origin at an angle of 60° with respect to the point $(x, y) = (1, 1)$, calculate its initial velocity (*m/s*). Herein, assume that $g = 10 \ m/s^2$.

Solution

As we know, the trajectory equation of a projectile is as follows.

$$y = \frac{-gx^2}{2v_0^2 \cos^2\theta} + x\tan\theta + y_0$$

Therefore:

$$1 = \frac{-10}{2v_0^2 \cos^2 60°} + \tan 60° + 0$$

$$\Rightarrow 1 = \frac{-10}{2v_0^2(\quad)^2} + (\quad)$$

$$\Rightarrow v_0^2 = (\quad)$$

$$\Rightarrow v_0 = 5.22 \ m/s$$

3.5 Power

3.18. The instantaneous position equation of a 2 kg object in the SI system is $x(t) = -t^2 + 4t^3$. Calculate the instantaneous power (W) of the object at the first second.

Difficulty level ○ Easy ● Normal ○ Hard

Calculation amount ○ Small ● Normal ○ Large

1) 122 W
2) 292 W
3) 584 W
4) 440 W

References

1. Rahmani-Andebili, M. (2023). Calculus III – Practice Problems, Methods, and Solutions, Springer Nature.
2. Rahmani-Andebili, M. (2023). Calculus II – Practice Problems, Methods, and Solutions, Springer Nature.
3. Rahmani-Andebili, M. (2023). Calculus I (2nd Ed.) – Practice Problems, Methods, and Solutions, Springer Nature.
4. Rahmani-Andebili, M. (2024). Precalculus (2nd Ed.) – Practice Problems, Methods, and Solutions, Springer Nature.

Linear Kinematics: Part B

4

Abstract

In this chapter, the problems of the third chapter are fully solved, in detail, step-by-step, and with different methods.

4.1 Average Velocity

4.1. The velocity-time curve of the object is shown in Fig. 4.1. As we know, the average velocity of an object can be calculated as follows.

$$\bar{v} = \frac{\Delta r}{\Delta t}$$

Moreover, the area under the velocity-time curve of an object is its traveled distance [1–4].

Thus:

$$\bar{v} = \frac{Area\ under\ v - t\ curve}{Total\ time}$$

$$\Rightarrow \bar{v} = \frac{\frac{1}{2}(30 + 10)30}{30}$$

$$\Rightarrow \bar{v} = 20\ m/s$$

Choice (4) is the answer.

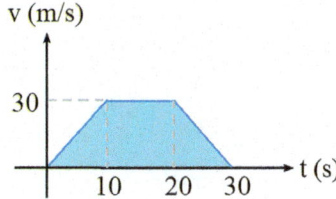

Fig. 4.1 The figure concerned with Problem 4.1

© The Author(s), under exclusive license to Springer Nature Switzerland AG 2025
M. Rahmani-Andebili, *General Physics I*, https://doi.org/10.1007/978-3-031-92862-8_4

4.2. The velocity-time curves of the objects are shown in Fig. 4.2. As we know, the average velocity of an object can be calculated as follows.

$$\bar{v} = \frac{\Delta r}{\Delta t}$$

Moreover, the area under the velocity-time curve of an object is its traveled distance.

The problem can be solved as follows:

$$\bar{v_A} = \frac{\Delta r}{\Delta t} = \frac{Area\ under\ v - t\ curve}{\Delta t} = \frac{S_A}{\Delta t}$$

$$\bar{v_B} = \frac{\Delta r}{\Delta t} = \frac{Area\ under\ v - t\ curve}{\Delta t} = \frac{S_B}{\Delta t}$$

Since $S_A > S_B$, we can conclude that:

$$\bar{v_A} > \bar{v_B}$$

Choice (2) is the answer.

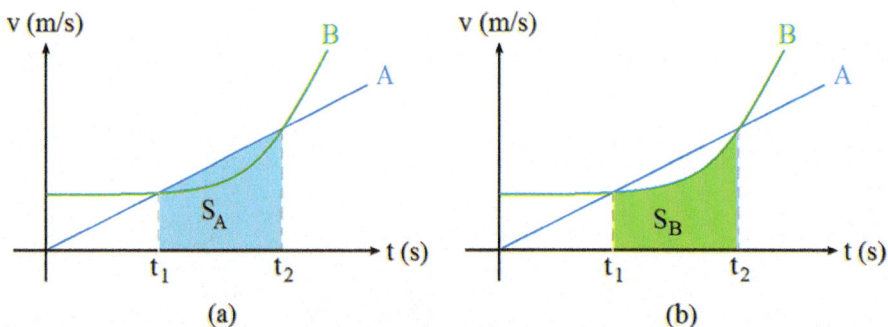

Fig. 4.2 The figure concerned with Problem 4.2

4.3. Based on the information given in the problem, we have:

$$(x_1, y_1) = (2, 5) \tag{4.1}$$

$$(x_2, y_2) = (5, 9) \tag{4.2}$$

$$\Delta t = 2 \tag{4.3}$$

The problem can be solved as follows:

$$\bar{v} = \frac{\Delta r}{\Delta t} = \frac{\sqrt{(x_2 - x_1)^2 + (y_2 - y_1)^2}}{\Delta t}$$

$$\Rightarrow \bar{v} = \frac{\sqrt{(5-2)^2 + (9-5)^2}}{2} = \frac{\sqrt{3^2 + 4^2}}{2}$$

$$\Rightarrow \bar{v} = 2.5 \, \frac{m}{s}$$

Choice (3) is the answer.

─────────────────── ■ ■ ■ ───────────────────

4.4. Based on the information given in the problem, we have:

$$v_1 = v, \quad \Delta t_1 = t \tag{4.4}$$

$$v_2 = 2v, \quad \Delta t_2 = 2t \tag{4.5}$$

$$v_3 = 3v, \quad \Delta t_3 = 3t \tag{4.6}$$

The problem can be solved as follows:

$$\bar{v} = \frac{\Delta r}{\Delta t} = \frac{\Delta r_1 + \ldots + \Delta r_N}{\Delta t_1 + \ldots + \Delta t_N}$$

$$\Rightarrow \bar{v} = \frac{v_1 \Delta t_1 + \ldots + v_N \Delta t_N}{\Delta t_1 + \ldots + \Delta t_N}$$

For this problem, we have:

$$\bar{v} = \frac{vt + 4vt + 9vt}{t + 2t + 3t} = \frac{14vt}{6t}$$

$$\Rightarrow \bar{v} = \frac{7}{3} v$$

Choice (4) is the answer.

─────────────────── ■ ■ ■ ───────────────────

4.5. Based on the information given in the problem, we have:

$$v_1 = v, \quad \Delta x_1 = x \tag{4.7}$$

$$v_2 = 2v, \quad \Delta x_2 = 2x \tag{4.8}$$

$$v_3 = 5v, \quad \Delta x_3 = 5x \tag{4.9}$$

The problem can be solved as follows:

$$\bar{v} = \frac{\Delta r}{\Delta t} = \frac{\Delta r_1 + \ldots + \Delta r_N}{\Delta t_1 + \ldots + \Delta t_N}$$

$$\Rightarrow \bar{v} = \frac{\Delta r_1 + \ldots + \Delta r_N}{\frac{\Delta x_1}{v_1} + \ldots + \frac{\Delta x_N}{v_N}}$$

For this problem, we have:

$$\bar{v} = \frac{x + 2x + 5x}{\frac{x}{v} + \frac{2x}{2v} + \frac{5x}{5v}} = \frac{8x}{3\frac{x}{v}}$$

$$\Rightarrow \bar{v} = \frac{8}{3}v$$

Choice (3) is the answer.

4.6. Based on the information given in the problem, we have:

$$y = \sqrt{3}x + 1 \tag{4.10}$$

$$v = 20\frac{m}{s} \tag{4.11}$$

$$\Delta t = 3 \tag{4.12}$$

From (4.10), it is noticed that the slope of the line is $\sqrt{3}$. Thus:

$$\tan\theta = \sqrt{3} \Rightarrow \theta = 60° \tag{4.13}$$

The problem can be decomposed and solved only on x-axis as follows:

$$v_x = v\cos\theta \tag{4.14}$$

Solving (4.13) and (4.14):

$$v_x = 20\cos 60° = 20 \times \frac{1}{2} = 10 \; m/s$$

$$\Rightarrow \Delta x = v_x\Delta t = 10 \times 3$$

$$\Rightarrow \Delta x = 30 \; m$$

Choice (1) is the answer.

Notes

In this problem, the relations below have been used.

$$Slope = \tan\theta$$

$$v_x = v\cos\theta$$

$$v_y = v\sin\theta$$

$$v_x = \frac{\Delta x}{\Delta t}$$

$$v_y = \frac{\Delta y}{\Delta t}$$

$$\tan^{-1}\sqrt{3} = 60°$$

$$\cos 60° = \frac{1}{2}$$

4.2   Relative Motion

4.7. Based on the information given in the problem, we have:

$$v_1 = 10\ \frac{m}{s}$$

$$v_2 = 20\ \frac{m}{s}$$

$$y = 300\ m$$

The problem can be solved as follows:

$$\Delta y = v_{relative}\Delta t$$

$$\Rightarrow 300 = (10 + 20)\Delta t$$

$$\Rightarrow \Delta t = 10\ s$$

Choice (2) is the answer.

4.3   Free Fall and Uniformly Accelerated Motion

4.8. Based on the information given in the problem, we have:

$$v_x = 450 \ m/s$$

$$x = 45 \ m$$

$$g = 10 \ m/s^2$$

The acceleration of the bullet in the x-axis is zero. Thus:

$$\Delta x = v_x \Delta t \Rightarrow 45 = 450 \times \Delta t \Rightarrow \Delta t = 0.1 \ s$$

The bullet will fall about Δy during the Δt that can be calculated as follows.

$$\Delta y = -\frac{1}{2}gt^2$$

$$\Rightarrow \Delta y = -\frac{1}{2} \times 10 \times 0.1^2 = -0.05 \ m$$

$$\Rightarrow \Delta y = -5 \ cm$$

Therefore, the target point must be 5 cm above the real target point. Choice (1) is the answer.

> **Notes**
>
> In this problem, the relations below have been used.
>
> $$x = v_x t + x_0 \Leftrightarrow \Delta x = v_x \Delta t$$
>
> $$y = -\frac{1}{2}gt^2 + v_{y0}t + y_0 \Leftrightarrow \Delta y = -\frac{1}{2}gt^2 + v_{y0}t$$

■ ■ ■

4.4   Projectile Motion

4.9. Based on the information given in the problem, we have:

$$\theta_1 = 30°$$

$$\theta_2 = 60°$$

The range of a projectile can be calculated as follows.

$$R = \frac{v_0^2}{g} \sin 2\theta$$

Therefore:

$$\frac{R_2}{R_1} = \frac{\dfrac{v_0^2}{g} \sin(2 \times 60°)}{\dfrac{v_0^2}{g} \sin(2 \times 30°)} = \frac{\sin(120°)}{\sin(60°)} = \frac{\dfrac{\sqrt{3}}{2}}{\dfrac{\sqrt{3}}{2}}$$

$$\Rightarrow \frac{R_2}{R_1} = 1$$

Choice (2) is the answer.

> **Notes**
>
> In this problem, the relations below have been used.
>
> $$\sin(120°) = \frac{\sqrt{3}}{2}$$
>
> $$\sin(60°) = \frac{\sqrt{3}}{2}$$

4.10. Based on the information given in the problem, we have:

$$\theta_1 = 30°$$

$$\theta_2 = 60°$$

The heigh of the peak point (apex) of a projectile can be calculated as follows.

$$h = \frac{v_0^2 \sin^2\theta}{2g}$$

Therefore:

$$\frac{h_2}{h_1} = \frac{\dfrac{v_0^2 \sin^2 60°}{2g}}{\dfrac{v_0^2 \sin^2 30°}{2g}} = \frac{\sin^2 60°}{\sin^2 30°} = \frac{\left(\dfrac{\sqrt{3}}{2}\right)^2}{\left(\dfrac{1}{2}\right)^2} = \frac{\dfrac{3}{4}}{\dfrac{1}{4}}$$

$$\Rightarrow \frac{h_2}{h_1} = 3$$

Choice (1) is the answer.

> **Notes**
>
> In this problem, the relations below have been used.
>
> $$\sin(60°) = \frac{\sqrt{3}}{2}$$
>
> $$\sin(30°) = \frac{1}{2}$$

4.11. As we know, the height of the peak point of a projectile can be calculated as follows.

$$h = \frac{v_0^2 \sin^2\alpha}{2g}$$

As can be seen, the value of $\sin^2\alpha$ is maximum if $\alpha = 90°$.

Choice (4) is the answer.

4.12. As we know, the range of a projectile can be calculated as follows.

$$R = \frac{v_0^2 \sin 2\alpha}{g}$$

As can be seen, the value of $\sin 2\alpha$ is maximum if $\alpha = 45°$.

Choice (3) is the answer.

4.13. Based on the information given in the problem, we have:

$$v_0 = 2v_{peak}$$

As we know, the velocity at peak point is equal to v_{0x} because $v_y = 0$ and $v_x = v_{0x} =$ Constant. Therefore:

$$v_0 = 2v_{0x}$$

At the time that the projection occurs, we have:

$$\cos \alpha = \frac{v_{0x}}{v_0} \Rightarrow \cos \alpha = \frac{v_{0x}}{2v_{0x}} = \frac{1}{2}$$

$$\Rightarrow \alpha = 60^\circ$$

Choice (2) is the answer.

Notes

In this problem, the relations below have been used.

$$v_x = v \cos \theta$$

$$\cos 60^\circ = \frac{1}{2}$$

■ ■ ■

4.14. Based on the information given in the problem, we have:

$$v_0 = 30 \ m/s$$

$$v_{peak} = 10 \ m/s$$

$$g = 10 \ m/s^2$$

As we know, the velocity at peak point is equal to v_{0x} because $v_y = 0$ and $v_x = v_{0x} =$ Constant. Therefore:

$$v_{0x} = 10 \ m/s$$

The y-component of the initial velocity can be calculated as follows.

$$v_0^2 = v_{0x}^2 + v_{0y}^2 \Rightarrow 30^2 = 10^2 + v_{0y}^2 \Rightarrow v_{0y}^2 = 800$$

The heigh of the apex can be calculated as follows.

$$v_y^2 - v_{0y}^2 = -2gy \Rightarrow 0 - 800 = -20h$$

$$\Rightarrow h = 40 \ m$$

Choice (2) is the answer.

■ ■ ■

4.15. Based on the information given in the problem, we have:

$$K_{peak} = 0.75K_0$$

As we know, the velocity at peak point is equal to v_{0x} because $v_y = 0$ and $v_x = v_{0x} = $ Constant. Hence:

$$\frac{1}{2}mv_{0x}^2 = 0.75\left(\frac{1}{2}mv_0^2\right) \Rightarrow v_{0x}^2 = 0.75v_0^2$$

$$\Rightarrow (v_0\cos\alpha)^2 = \frac{3}{4}v_0^2 \Rightarrow \cos^2\alpha = \frac{3}{4}$$

$$\Rightarrow \cos\alpha = \frac{\sqrt{3}}{2} \Rightarrow \alpha = \frac{\pi}{6} \ rad$$

Choice (1) is the answer.

> **Notes**
>
> In this problem, the relations below have been used.
>
> $$K = \frac{1}{2}mv^2$$
>
> $$v_x = v\cos\theta$$
>
> $$\cos\frac{\pi}{6} = \frac{\sqrt{3}}{2}$$

4.16. Based on the information given in the problem, we have:

$$\alpha = 30°$$

$$K_0 = 60 \ J$$

As we know, the velocity at peak point is equal to v_{0x} because $v_y = 0$ and $v_x = v_{0x} = $ Constant. Hence:

$$\frac{1}{2}mv_0^2 = 60$$

The kinetic energy at the peak point can be calculated as follows.

$$K_{peak} = \frac{1}{2}mv_{0x}^2 = \frac{1}{2}m(v_0\cos\alpha)^2 = \left(\frac{1}{2}mv_0^2\right)(\cos^2\alpha)$$

Therefore:

$$K_{peak} = 60\cos^2 30° = 60 \times \frac{3}{4}$$

$$\Rightarrow K_{peak} = 45 \, J$$

Choice (3) is the answer.

Notes

In this problem, the relations below have been used.

$$K = \frac{1}{2}mv^2$$

$$v_x = v\cos\theta$$

$$\cos\frac{\pi}{6} = \frac{\sqrt{3}}{2}$$

■ ■ ■

4.17. Based on the information given in the problem, we have:

$$\theta = 45°$$

$$(x, y) = (24, 14)$$

$$y_0 = 0$$

$$g = 10 \, m/s^2$$

As we know, the trajectory equation of a projectile is as follows.

$$y = \frac{-gx^2}{2v_0^2\cos^2\theta} + x\tan\theta + y_0$$

Therefore:

$$14 = \frac{-10 \times 24^2}{2v_0^2\cos^2 45°} + 24\tan 45° + 0 \Rightarrow 14 = \frac{-10 \times 24^2}{2v_0^2 \times \frac{1}{2}} + 24 \times 1$$

$$\Rightarrow -10 = \frac{-10 \times 24^2}{v_0^2} \Rightarrow v_0^2 = 24^2$$

$$\Rightarrow v_0 = 24 \, m/s$$

Choice (3) is the answer.

Notes

In this problem, the relations below have been used.

$$\cos 45° = \frac{\sqrt{2}}{2}$$

$$\tan 45° = 1$$

4.5   Power

4.18. Based on the information given in the problem, we have:

$$m = 2\ kg$$

$$x(t) = -t^2 + 4t^3$$

$$t = 1\ s$$

The velocity and acceleration of the object can be calculated as follows.

$$v(t) = \frac{dx(t)}{dt} = -2t + 12t^2$$

$$a(t) = \frac{dv(t)}{dt} = \frac{d^2x(t)}{dt^2} = -2 + 24t$$

The instantaneous power of the object can be calculated as follows.

$$p(t) = F(t)v(t) = ma(t)v(t)$$

$$\Rightarrow p(1) = ma(1)v(1) = 2(-2 + 24)(-2 + 12)$$

$$\Rightarrow p = 440\ W$$

Choice (4) is the answer.

Notes

In this problem, the relations below have been used.

$$v = \frac{dx}{dt}$$

$$a = \frac{d^2x}{dt^2}$$

$$p = Fv$$

$$F = ma$$

References

1. Rahmani-Andebili, M. (2023). Calculus III – Practice Problems, Methods, and Solutions, Springer Nature.
2. Rahmani-Andebili, M. (2023). Calculus II – Practice Problems, Methods, and Solutions, Springer Nature.
3. Rahmani-Andebili, M. (2023). Calculus I (2nd Ed.) – Practice Problems, Methods, and Solutions, Springer Nature.
4. Rahmani-Andebili, M. (2024). Precalculus (2nd Ed.) – Practice Problems, Methods, and Solutions, Springer Nature.

Linear Dynamics: Part A

5

Abstract

In this chapter, the basic and advanced problems of linear dynamics are studied. The subjects include Newton's laws, Newton's laws on inclined surface, Newton's Laws in an elevator, centripetal force, momentum, position-dependent force, potential energy, conservative force, work-energy theorem, conservation of mechanical energy principle, and spring force and Hooke's law. Herein, different types of problems and exercises are presented that are categorized as follows.

- ***Problems with detailed solution***: They have been designed to teach students the subjects in detail. Moreover, they have been categorized in different levels based on their difficulty levels (easy, normal, and hard) and calculation amounts (small, normal, and large).
- ***Partially solved exercises***: They have been designed to encourage students to practice problems while guiding them through the problem-solving procedure and hinting the required formulas.
- ***Exercises with final answer***: They have been designed to encourage students to practice more by themselves while hinting them by the final answer as well as to help instructors to give tests or quizzes.

5.1 Newton's Laws

Problem

5.1. In the system illustrated in Fig. 5.1, calculate the normal force that the wall exerts on the sphere if the tension force of the string is 200 N [1–4]. Herein, assume that the surface is frictionless and $\sin 53° = \frac{4}{5}$.

Difficulty level   ○ Easy   ● Normal   ○ Hard

Calculation amount   ● Small   ○ Normal   ○ Large

1) 120 N
2) 160 N
3) 240 N
4) 200 N

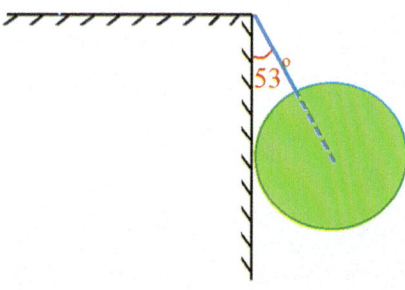

Fig. 5.1 The figure concerned with Problem 5.1

Problem

5.2. The mass of the box shown in Fig. 5.2 is 10 kg. Calculate the tension force of the second string (N). Herein, assume that $g = 10 \ m/s^2$.

Difficulty level   ○ Easy   ● Normal   ○ Hard
Calculation amount   ○ Small   ● Normal   ○ Large

1) $T_2 = \dfrac{200}{1 + \sqrt{3}}$

2) $T_2 = \dfrac{100}{1 + \sqrt{3}}$

3) $T_2 = 200\left(1 + \sqrt{3}\right)$

4) $T_2 = \dfrac{200}{1 + \sqrt{2}}$

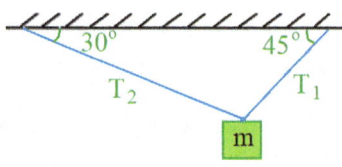

Fig. 5.2 The figure concerned with Problem 5.2

Exercise

In the previous problem, calculate the tension force of the first string (N).

Final Answer

$$T_1 = \frac{100\left(3 - \sqrt{3}\right)}{\sqrt{2}} \ N$$

Problem

5.3. In the system shown in Fig. 5.3, $m_1 = 2 \ kg$ and $m_2 = 1 \ kg$. Calculate the acceleration of the system (m/s^2). Herein, assume that $g = 10 \ m/s^2$.

Difficulty level   ○ Easy   ● Normal   ○ Hard
Calculation amount   ○ Small   ● Normal   ○ Large

1) $a = \dfrac{10}{3}$

2) $a = \dfrac{20}{3}$

3) $a = 1$

4) $a = \dfrac{40}{3}$

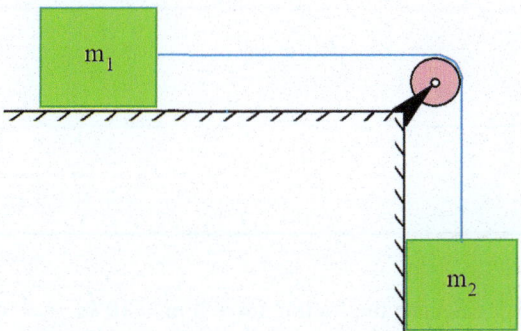

Fig. 5.3 The figure concerned with Problem 5.3

Exercise

In the previous problem, calculate the string tension force (N).

Final Answer

$T = \dfrac{20}{3} N$

Problem

5.4. In the system shown in Fig. 5.4, the surface is frictionless and $T_3 = 60\ N$. Calculate the acceleration of the system and T_1 if $m_1 = 10\ kg$, $m_2 = 20\ kg$, and $m_3 = 30\ kg$.

Difficulty level ○ Easy ● Normal ○ Hard
Calculation amount ○ Small ● Normal ○ Large

1) $a = 2\ m/s^2$, $T_1 = 10\ N$
2) $a = 1\ m/s^2$, $T_1 = 30\ N$
3) $a = 1\ m/s^2$, $T_1 = 10\ N$
4) $a = 2\ m/s^2$, $T_1 = 30\ N$

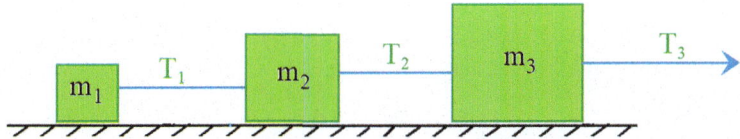

Fig. 5.4 The figure concerned with Problem 5.4

Exercise

In the previous problem, calculate T_2.

Final Answer

$T_2 = 30\ N$

Problem

5.5. In the system illustrated in Fig. 5.5, calculate the friction force if $m = 10\ kg$, $\mu_s = 0.5$, $\mu_k = 0.4$, and $F = 30\ N$. Herein, assume that $g = 10\ m/s^2$.

Difficulty level   ○ Easy   ○ Normal   ● Hard
Calculation amount   ○ Small   ● Normal   ○ Large

1) $100\ N$
2) $30\ N$
3) $50\ N$
4) $60\ N$

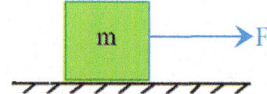

Fig. 5.5 The figure concerned with Problem 5.5

Exercise

In the previous problem, calculate the friction force for $F > 50\ N$.

Final Answer

$f_k = 40\ N$

Problem

5.6. As is illustrated in Fig. 5.6, the horizontal force $53\ N$ is exerted on the $22\ N$ weight box. Determine if the box will move down if the static friction constant between the box and wall is 0.6.

Difficulty level   ○ Easy   ○ Normal   ● Hard
Calculation amount   ○ Small   ● Normal   ○ Large

1) No
2) Yes
3) The data is incomplete to answer the question.
4) It moves first, and then stops.

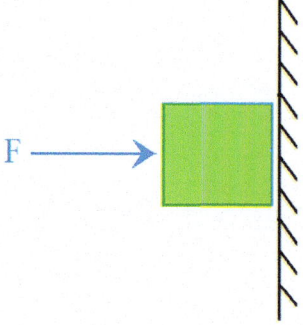

Fig. 5.6 The figure concerned with Problem 5.6

Problem

5.7. In the system shown in Fig. 5.7, the horizontal surface is frictionless; however, the static friction constant between the two masses is 0.25. Calculate the minimum force F so that the small mass does not move downward. Herein, $m_2 = 2\ kg$, $m_1 = 0.5\ kg$, and $g = 10\ m/s^2$.

Difficulty level   ○ Easy   ○ Normal   ● Hard
Calculation amount   ○ Small   ● Normal   ○ Large

1) 25 N
2) 60 N
3) 50 N
4) 100 N

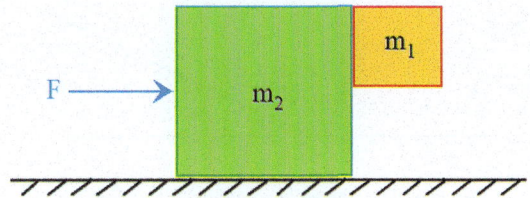

Fig. 5.7 The figure concerned with Problem 5.7

5.2   Newton's Laws on Inclined Surface

Problem

5.8. As is shown in Fig. 5.8, an object on an inclined surface moves down with a constant velocity. Calcucate the kinetic friction constant.

Difficulty level   ○ Easy   ● Normal   ○ Hard
Calculation amount   ○ Small   ● Normal   ○ Large

1) $\dfrac{\sqrt{3}}{3}$
2) $\dfrac{\sqrt{3}}{2}$
3) $\sqrt{3}$
4) 1

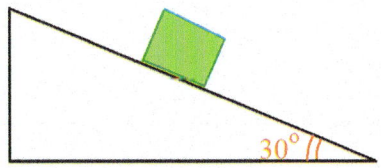

Fig. 5.8 The figure concerned with Problem 5.11

Problem

5.9. The inclined surface shown in Fig. 5.9 is frictionless and the mass of the object is 5 *kg*. Calculate the tension force of the rope. Herein, assume that $g = 10 \ m/s^2$.

Difficulty level   ○ Easy   ● Normal   ○ Hard
Calculation amount   ○ Small   ● Normal   ○ Large

1) $T = 50 \ N$
2) $T = 25\sqrt{3} \ N$
3) $T = 25 \ N$
4) $T = 50\sqrt{3} \ N$

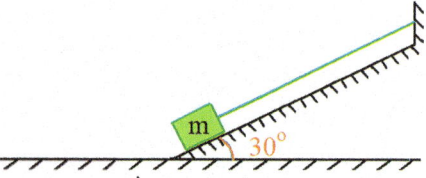

Fig. 5.9 The figure concerned with Problem 5.8

Partially Solved Exercise

In the previous problem, calculate the normal force of the inclined surface (N).

Solution

By applying Newton's second law on the y-axis, we have:

$$\sum F_y = ma_y$$

$$\Rightarrow (\qquad) - (\qquad) = (\qquad) \Rightarrow N = (\quad) \times (\quad) \times (\quad)$$

$$\Rightarrow N = 25\sqrt{3}$$

Notes

In this problem, the relation below is needed.

$$\cos 30° = \frac{\sqrt{3}}{2}$$

Problem

5.10. In Problem 5.9, assume that the rope is suddenly cut. Calculate the magnitude of the acceleration of the object (m/s^2).

Difficulty level   ○ Easy   ● Normal   ○ Hard
Calculation amount   ○ Small   ● Normal   ○ Large

1) 0
2) 1
3) 3
4) 5

Partially Solved Exercise

The system illustrated in Fig. 5.10 is in the equilibrium state. Calculate the value of $\frac{m_2}{m_1}$.

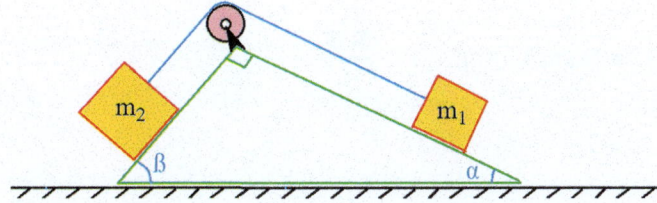

Fig. 5.10   The figure concerned with the partially solved exercise

Solution

Figure 5.11 shows all the forces applied on the objects. Now, we can write Newton's second law on the x- and y-axis for each mass. Since the system is in the equilibrium state, $a_x = a_y = 0$

Newton's second law on the x-axis of m_1:

$$\sum F_x = m_1 a_x \Rightarrow (\qquad) - (\qquad) = (\qquad) \Rightarrow m_1 g \sin \alpha = T \qquad (5.1)$$

Newton's second law on the x-axis of m_2:

$$\sum F_y = m_2 a_x \Rightarrow (\qquad) - (\qquad) = (\qquad) \Rightarrow m_2 g \sin \beta = T \qquad (5.2)$$

Solving (5.1) and (5.2):

$$\frac{(\qquad)}{(\qquad)} = 1 \Rightarrow \frac{m_2}{m_1} = \frac{(\qquad)}{(\qquad)}$$

$$\Rightarrow \frac{m_2}{m_1} = \tan \alpha$$

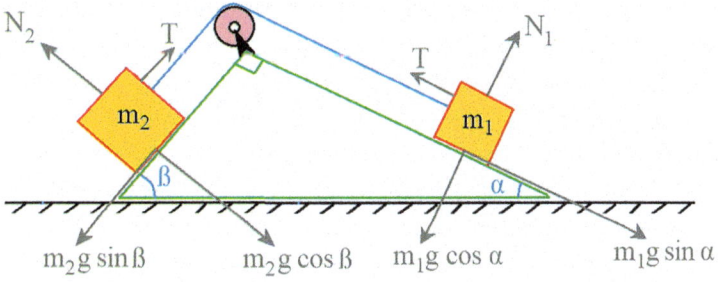

Fig. 5.11   The figure concerned with the partially solved exercise

Problem

5.11. As is illustrated in Fig. 5.12, an object which is located on a frictionless inclined surface is shot upward with the initial velocity v_0. Determine the distance that it will move on the inclined surface. Herein, α is the angle of the inclined surface.

Difficulty level ○ Easy ○ Normal ● Hard
Calculation amount ○ Small ● Normal ○ Large

1) $\dfrac{v_0}{g \sin \alpha}$

2) $\dfrac{v_0}{2g \sin \alpha}$

3) $\dfrac{v_0^2}{2g \sin \alpha}$

4) $\dfrac{v_0^2}{g \sin \alpha}$

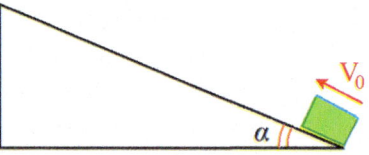

Fig. 5.12 The figure concerned with Problem 5.10

Exercise

In the previous problem, calculate the time that will take the object to move the distance on the inclined surface.

Final Answer

$t = \dfrac{v_0}{g \sin \alpha}$

5.3   Newton's Laws in an Elevator

Problem

5.12. In the system shown in Fig. 5.13, the mass m has been connected to an elevator through two weightless strings. Calculate the tension force of the right-hand side string (T_1) if the elevator goes up with the positive acceleration a.

Difficulty level ○ Easy ○ Normal ● Hard
Calculation amount ○ Small ○ Normal ● Large

1) $\left(\sqrt{3}+1\right)m(g+a)$

2) $\left(\sqrt{3}-1\right)m(g+a)$

3) $\sqrt{\dfrac{3}{2}}\left(\sqrt{3}+1\right)m(g+a)$

4) $\sqrt{\dfrac{3}{2}}\left(\sqrt{3}-1\right)m(g+a)$

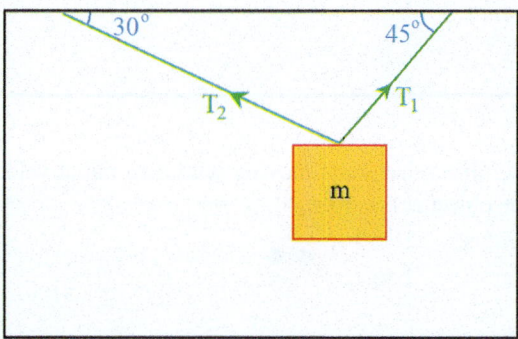

Fig. 5.13 The figure concerned with Problem 5.12

5.4 Centripetal Force

Problem

5.13. What must be the minimum speed of a car to not fall from the wall in the wall of death game? Herein, R is the radius of the cylindrical wall (Fig. 5.14).

Difficulty level   ○ Easy   ● Normal   ○ Hard
Calculation amount   ○ Small   ● Normal   ○ Large

1) $\sqrt{\dfrac{g}{R\mu_s}}$

2) $\sqrt{\mu_s R g}$

3) $\sqrt{\dfrac{\mu_s g}{R}}$

4) $\sqrt{\dfrac{Rg}{\mu_s}}$

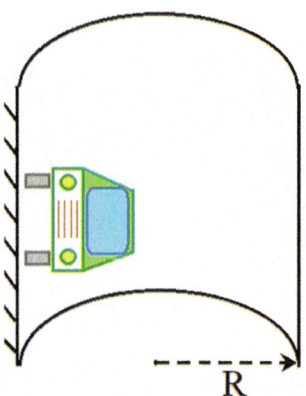

Fig. 5.14 The figure concerned with Problem 5.13

5.5   Momentum

Problem

5.14. The force-time curve of a 60 *kg* object that starts moving with zero initial velocity at the origin of time is shown in Fig. 5.15. Calculate the velocity of the object at $t = 30\ s$.

Difficulty level   ● Easy   ○ Normal   ○ Hard
Calculation amount   ● Small   ○ Normal   ○ Large

1) 5 *m/s*
2) 15 *m/s*
3) 20 *m/s*
4) 10 *m/s*

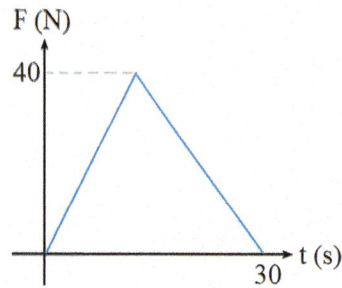

Fig. 5.15   The figure concerned with Problem 5.14

Exercise

Solve the previous problem while considering Fig. 5.16 as the force-time curve of the object.

Final Answer

$v_2 \approx 3.33\ m/s$

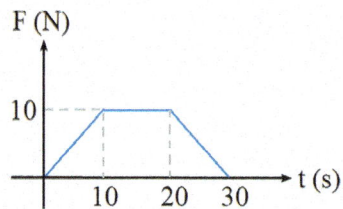

Fig. 5.16   The figure concerned with the exercise

5.6 Position-Dependent Force and Potential Energy

Problem

5.15. Calculate the amount of work done on an object under the force below to move it from $x = 1$ m to $x = 2$ m.

$$F(x) = -x + \frac{1}{x^3}$$

Difficulty level ● Easy ○ Normal ○ Hard
Calculation amount ○ Small ● Normal ○ Large
1) $-\frac{3}{8}$ J
2) $-\frac{9}{8}$ J
3) $-\frac{3}{2}$ J
4) $\frac{3}{8}$ J

Problem

5.16. The force equation of a spring in the SI system is $F(x) = -3x^2 - 8x^3$. Calculate the potential energy of the spring at $x = 1$ m if its potential energy at $x = 0$ m is 1 J.

Difficulty level ● Easy ○ Normal ○ Hard
Calculation amount ○ Small ● Normal ○ Large
1) 1 J
2) 2 J
3) 4 J
4) 8 J

5.7 Conservative Force

Problem

5.17. For what values of the parameters a, b, and c, the force vector $\vec{F} = (ax + 2by^2)\hat{i} + cxy\hat{j}$ is a conservative force?

Difficulty level ○ Easy ● Normal ○ Hard
Calculation amount ○ Small ● Normal ○ Large
1) $a = 4b$
2) $a = b = c$
3) $b = 2c$
4) $c = 4b$

Exercise

Determine if the force below is conservative or not.

$$\vec{F} = \left(x + 2y^2\right)\widehat{i} + 4xy\widehat{j}$$

Final Answer

It is a conservative force because $\nabla \times \vec{F} = 0$.

5.8   Work-Energy Theorem

Problem

5.18. Calculate the amount of work done on a 2 *kg* object from zero to one second that its equation of motion is $\vec{r} = (3t + 5t^3)\widehat{i}\ m/s$.

Difficulty level   ○ Easy   ● Normal   ○ Hard
Calculation amount   ○ Small   ● Normal   ○ Large
1) 315 *J*
2) 324 *J*
3) 480 *J*
4) 1080 *J*

Exercise

Calculate the amount of work done on a 1 *kg* object to increase its velocity from 1 *m/s* to 2 *m/s*.

Final Answer

$W = 1.5\ J$

5.9   Conservation of Mechanical Energy Principle

Problem

5.19. The height of the chair in the swing chair game changes from 0.5 *m* to 2 *m*. Calculate its maximum speed. Herein, assume that $g = 10\ m/s^2$.

Difficulty level   ○ Easy   ● Normal   ○ Hard
Calculation amount   ○ Small   ● Normal   ○ Large
1) 5.4 *m/s*
2) 7.7 *m/s*
3) 29.4 *m/s*
4) 35.2 *m/s*

5.10 Spring Force and Hooke's Law

Problem

5.20. A spring with the stiffness constant of 6 *N/m* is halved, and then connected in parallel as is shown in Fig. 5.17. Calculate the mass of the object (*kg*) if the system oscillates with the frequency of 3 *Hz*.

Difficulty level   ○ Easy   ● Normal   ○ Hard

Calculation amount   ○ Small   ● Normal   ○ Large

1) $\dfrac{1}{12\pi^2}$

2) $\dfrac{1}{6\pi^2}$

3) $\dfrac{1}{3\pi^2}$

4) $\dfrac{2}{3\pi^2}$

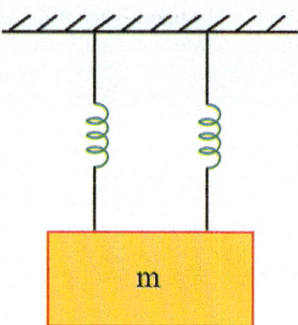

Fig. 5.17 The figure concerned with Problem 5.20

Exercise

Calculate the angular frequency of the oscillations of the system shown in Fig. 5.18.

Final Answer

$$\omega = \sqrt{\frac{2k}{3m}}$$

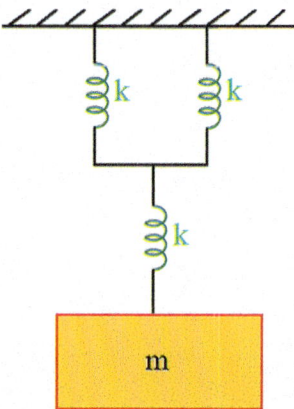

Fig. 5.18 The figure concerned with the exercise

Notes

In this problem, the relations below are needed.

The equivalent stiffness constant of two parallel springs can be calculated as follows.

$$k_{tot} = k_1 + k_2$$

The equivalent stiffness constant of two series springs can be calculated as follows.

$$k_{tot} = \frac{k_1 k_2}{k_1 + k_2}$$

References

1. Rahmani-Andebili, M. (2023). Calculus III – Practice Problems, Methods, and Solutions, Springer Nature.
2. Rahmani-Andebili, M. (2023). Calculus II – Practice Problems, Methods, and Solutions, Springer Nature.
3. Rahmani-Andebili, M. (2023). Calculus I (2nd Ed.) – Practice Problems, Methods, and Solutions, Springer Nature.
4. Rahmani-Andebili, M. (2024). Precalculus (2nd Ed.) – Practice Problems, Methods, and Solutions, Springer Nature.

Abstract

In this chapter, the problems of the fifth chapter are fully solved, in detail, step-by-step, and with different methods.

6.1 Newton's Laws

6.1. Based on the information given in the problem, we have:

$$T = 200 \, N$$

$$\sin 53\,° = \frac{4}{5}$$

Figure 6.1 shows all the forces applied on the object. Now, we can write Newton's second law on the x-axis as follows. Since the system is in the equilibrium state, $a_x = 0$ [1–4].

$$\sum F_x = ma_x$$

$$\Rightarrow N - T \sin 53\,° = 0 \Rightarrow N = T \sin 53\,° = 200 \times \frac{4}{5} \Rightarrow N = 160 \, N$$

Choice (2) is the answer.

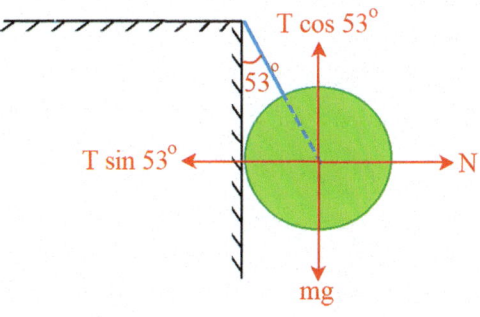

Fig. 6.1 The figure concerned with Problem 6.1

6.2. Based on the information given in the problem, we have:

$$m = 10 \; kg$$

$$g = 10 \; m/s^2$$

Figure 6.2b shows all the forces applied on the object. As can be seen, the tension force of the strings has been decomposed into their x- and y-components. Now, we can write Newton's second law on the x- and y-axis. Since the object is in the stationary state, the net force on each axis is zero.

Newton's second law on the x-axis:

$$\sum F_x = ma_x$$

$$\Rightarrow T_1 \cos 45^\circ - T_2 \cos 30^\circ = 0 \Rightarrow \frac{\sqrt{2}}{2} T_1 = \frac{\sqrt{3}}{2} T_2$$

$$\Rightarrow \sqrt{2} T_1 = \sqrt{3} T_2 \tag{6.1}$$

Newton's second law on the y-axis:

$$\sum F_y = ma_y$$

$$\Rightarrow T_1 \sin 45^\circ + T_2 \sin 30^\circ - mg = 0 \Rightarrow \frac{\sqrt{2}}{2} T_1 + \frac{1}{2} T_2 = mg$$

$$\Rightarrow \sqrt{2} T_1 + T_2 = 2mg = 2 \times 10 \times 10 \Rightarrow \sqrt{2} T_1 + T_2 = 200 \tag{6.2}$$

Solving (6.1) and (6.2):

$$\sqrt{3} T_2 + T_2 = 200 \Rightarrow T_2 = \frac{200}{1 + \sqrt{3}} \; N \tag{6.3}$$

Choice (1) is the answer.

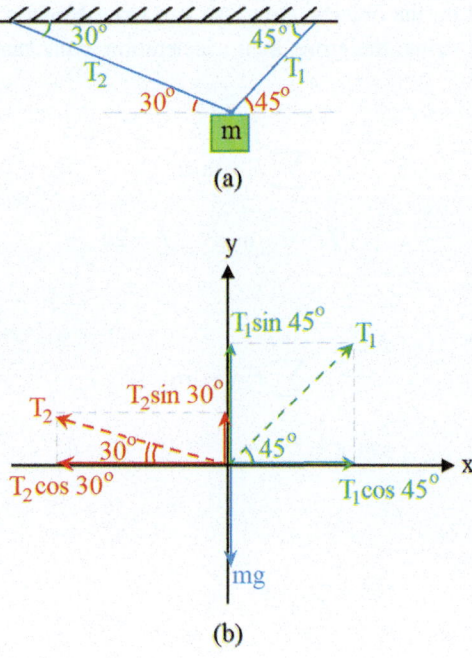

Fig. 6.2 The figure concerned with Problem 6.2

Notes

In this problem, the relations below have been used.

$$\cos 45° = \frac{\sqrt{2}}{2}$$

$$\cos 30° = \frac{\sqrt{3}}{2}$$

$$\sin 45° = \frac{\sqrt{2}}{2}$$

$$\sin 30° = \frac{1}{2}$$

6.3. Based on the information given in the problem, we have:

$$m_1 = 2\ kg$$

$$m_2 = 1\ kg$$

$$g = 10\ m/s^2$$

Figure 6.3 shows all the forces applied on the objects. Now, we can write Newton's second law for each mass. Also, the masses have been connected to a single string; therefore, their acceleration is the same. In other words, $a_1 = a_2 = a$.

Newton's second law on m_1:

$$\sum F = m_1 a_1$$

$$\Rightarrow T - 0 = m_1 a \Rightarrow T = 2a \tag{6.4}$$

Newton's second law on m_2:

$$\sum F = m_2 a_2$$

$$\Rightarrow m_2 g - T = m_2 a \Rightarrow 10 - T = a \tag{6.5}$$

Solving (6.4) and (6.5):

$$10 - 2a = a \Rightarrow a = \frac{10}{3} \ m/s^2 \tag{6.6}$$

Choice (1) is the answer.

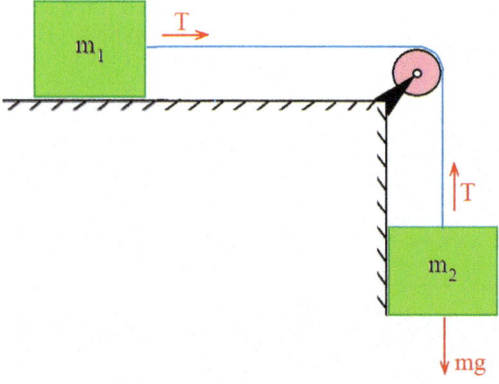

Fig. 6.3 The figure concerned with Problem 6.3

6.4. Based on the information given in the problem, we have:

$$T_3 = 60 \ N$$

$$m_1 = 10 \ kg$$

$$m_2 = 20 \ kg$$

$$m_3 = 30 \ kg$$

$$g = 10 \ m/s^2$$

Figure 6.4 shows all the forces applied on the objects. Now, we can apply Newton's second law. Also, since the masses have been connected to a single string, their acceleration is the same. In other words, $a_1 = a_2 = a_3 = a$.

Newton's second law for all masses:

$$\sum F = m_{total}a$$

$$\Rightarrow T_3 - 0 = (m_1 + m_2 + m_3)a \Rightarrow a = \frac{60}{10 + 20 + 30} \Rightarrow a = 1 \, m/s^2$$

Newton's second law for m_1:

$$\sum F = m_1 a_1$$

$$\Rightarrow T_1 - 0 = m_1 a \Rightarrow T_1 = 10 \times 1 \Rightarrow T_1 = 10 \, N$$

Choice (3) is the answer.

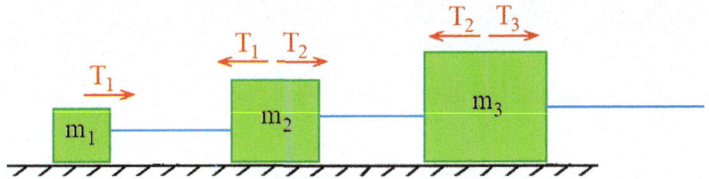

Fig. 6.4 The figure concerned with Problem 6.4

6.5. Based on the information given in the problem, we have:

$$m = 10 \, kg$$

$$\mu_s = 0.5$$

$$\mu_k = 0.4$$

$$F = 30 \, N$$

$$g = 10 \, m/s^2$$

Figure 6.5 shows all the forces applied on the objects. Now, we can write Newton's second law on the x- and y-axis.

Newton's second law on the y-axis:

$$\sum F_y = ma_y$$

$$N - mg = 0 \Rightarrow N = 10 \times 10 = 100 \, N \tag{6.7}$$

Newton's second law on the x-axis:

$$\sum F_x = ma_x$$

$$\Rightarrow F - f_k = ma \Rightarrow F - \mu_k N = ma \Rightarrow 30 - 0.4 \times 100 = 10a \Rightarrow a = -1\ m/s^2 \qquad (6.8)$$

As can be seen, the acceleration is negative that implies the object does not move at all. Thus, we need to consider zero acceleration in Newton's second law as follows.

$$\Rightarrow F - f_s = 0 \Rightarrow f_s = F \Rightarrow f_s = 30\ N$$

Choice (2) is the answer.

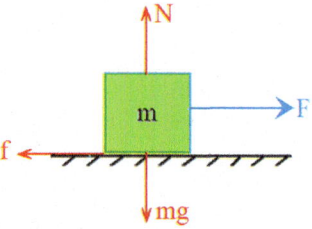

Fig. 6.5 The figure concerned with Problem 6.5

6.6. Based on the information given in the problem, we have:

$$F = 53\ N$$

$$W = 22\ N$$

$$\mu_s = 0.6$$

Figure 6.6 shows all the forces applied on the object. Now, we can apply Newton's second law as follows.

Newton's second law on the x-axis:

$$\sum F_x = ma_x$$

$$N - F = 0 \Rightarrow N = F = 53\ N$$

On the other hand, we have:

$$f_s = \mu_s N$$

$$\Rightarrow f_s = 0.6 \times 53 = 31.8\ N$$

As can be noticed, $f_s > W$; therefore, the object will never move down. Hence, the answer is no. Choice (1) is the answer.

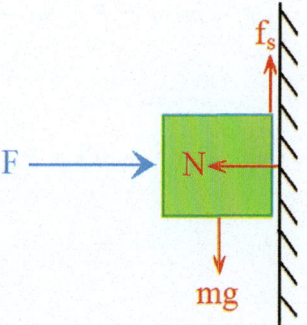

Fig. 6.6 The figure concerned with Problem 6.6

6.7. Based on the information given in the problem, we have:

$$\mu_s = 0.25$$

$$m_1 = 0.5 \; kg$$

$$m_2 = 2 \; kg$$

$$g = 10 \; m/s^2$$

The minimum force F that the small mass does not move downward happens when the net force exerted on the small mass is zero. Figure 6.7 shows all the forces applied on the object. Now, we can apply Newton's second law on each axis as follows.

Newton's second law on the x-axis for m_1:

$$\sum F_x = m_1 a_x$$

$$\Rightarrow N - 0 = m_1 a \Rightarrow N = m_1 a \tag{6.9}$$

Newton's second law on the y-axis for m_1:

$$\sum F_y = m_1 a_y$$

$$\Rightarrow m_1 g - f_s = 0 \Rightarrow m_1 g - \mu_s N = 0 \Rightarrow \mu_s N = m_1 g \tag{6.10}$$

Solving (6.9) and (6.10):

$$\mu_s m_1 a = m_1 g \Rightarrow a = \frac{g}{\mu_s} \tag{6.11}$$

On the other hand, since the small mass does not move downward, both masses move together. By applying Newton's second law for them on the x-axis, we have:

$$\sum F_x = (m_1 + m_2) a_x$$

$$F - 0 = (m_1 + m_2)a \Rightarrow F = (m_1 + m_2)a \tag{6.12}$$

Solving (6.11) and (6.12):

$$F = (m_1 + m_2)\frac{g}{\mu_s} = (0.5 + 2)\frac{10}{0.25}$$

$$F = 100\ N$$

Choice (4) is the answer.

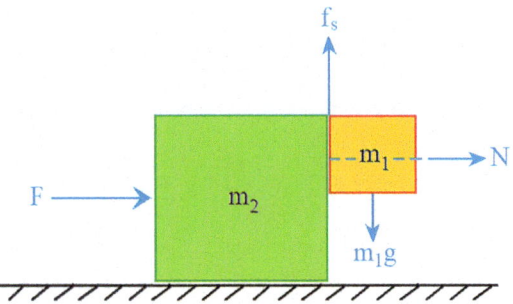

Fig. 6.7 The figure concerned with Problem 6.7

6.2   Newton's Laws on Inclined Surface

6.8. Figure 6.8 shows all the forces applied on the object. Now, we can write Newton's second law on the x- and y-axis. Since the object moves down with a constant velocity, its acceleration is zero.

Newton's second law on the x-axis:

$$\sum F_x = ma_x$$

$$\Rightarrow mg \sin \alpha - \mu_k N = 0 \Rightarrow mg \sin \alpha = \mu_k N \tag{6.13}$$

Newton's second law on the y-axis:

$$\sum F_y = ma_y$$

$$\Rightarrow N - mg \cos \alpha = 0 \Rightarrow N = mg \cos \alpha \tag{6.14}$$

Solving (6.13) and (6.14):

$$mg \sin \alpha = \mu_k mg \cos \alpha$$

$$\Rightarrow \sin \alpha = \mu_k \cos \alpha$$

$$\Rightarrow \mu_k = \tan \alpha$$

For $\alpha = 30°$:

$$\mu_k = \tan 30°$$

$$\Rightarrow \mu_k = \frac{\sqrt{3}}{3}$$

Choice (1) is the answer.

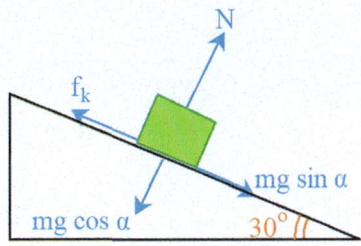

Fig. 6.8 The figure concerned with Problem 6.8

Notes

In this problem, the relations below have been used.

$$\tan \alpha = \frac{\sin \alpha}{\cos \alpha}$$

$$\tan 30° = \frac{\sqrt{3}}{3}$$

6.9. Based on the information given in the problem, we have:

$$m = 5 \ kg$$

$$g = 10 \ m/s^2$$

Figure 6.9 shows all the forces applied on the object. As can be seen, the weight of the mass has been decomposed into their x- and y-components. Now, we can write Newton's second law on the x-axis as follows. Since the object is in the stationary state, the net force on each axis is zero.

$$\sum F_x = ma_x$$

$$\Rightarrow T - mg\sin\theta = 0 \Rightarrow T = mg\sin\theta = 5 \times 10 \times \sin 30$$

$$\Rightarrow T = 25\ N$$

Choice (3) is the answer.

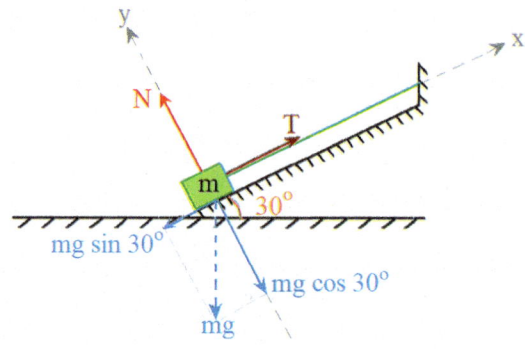

Fig. 6.9 The figure concerned with Problem 6.9

Notes

In this problem, the relation below has been used.

$$\sin 30^\circ = \frac{1}{2}$$

■ ■ ■

6.10. If the rope is suddenly cut, the object will accelerate downwards. Now, we can write Newton's second law on the x-axis as follows (Fig. 6.10).

$$\sum F_x = ma_x$$

$$\Rightarrow -mg\sin\theta = ma_x \Rightarrow a_x = -g\sin\theta = -10 \times \sin 30$$

$$\Rightarrow a = -5\ m/s^2 \Rightarrow |a| = 5\ m/s^2$$

Choice (4) is the answer.

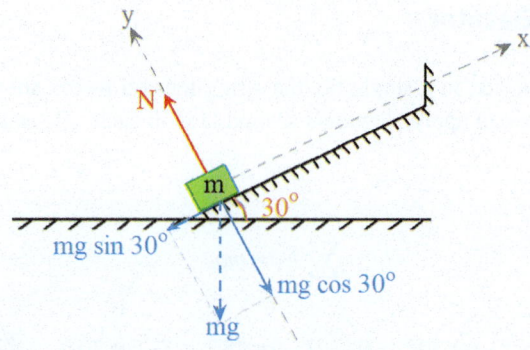

Fig. 6.10 The figure concerned with Problem 6.10

6.11. Figure 6.11 shows all the forces applied on the object. Now, we can apply Newton's second law on each axis as follows.

Newton's second law on the y-axis:

$$\sum F_y = ma_y$$

$$\Rightarrow N - mg\cos\alpha = 0 \Rightarrow N = mg\cos\alpha$$

Newton's second law on the x-axis:

$$\sum F_x = ma_x$$

$$\Rightarrow mg\sin\alpha = ma \Rightarrow a = g\sin\alpha$$

On the other hand, from linear kinematics, we can write:

$$v^2 - v_0^2 = -2ax$$

$$\Rightarrow 0 - v_0^2 = -2xg\sin\alpha$$

$$\Rightarrow x = \frac{v_0^2}{2g\sin\alpha}$$

Choice (3) is the answer.

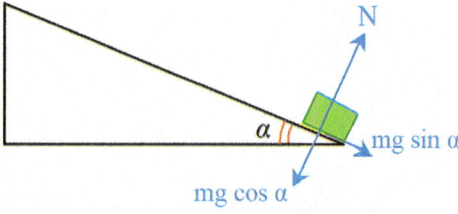

Fig. 6.11 The figure concerned with Problem 6.11

6.3   Newton's Laws in an Elevator

6.12. Since the object has been connected to a single tie, the string tension forces are different. Figure 6.12 shows all the forces applied on the object. Now, we can apply Newton's second law on each axis as follows.

Newton's second law on the x-axis:

$$\sum F_x = ma_x$$

$$\Rightarrow T_1 \cos 45° - T_2 \cos 30° = 0 \Rightarrow T_1 \cos 45° = T_2 \cos 30° \Rightarrow \frac{\sqrt{2}}{2} T_1 = \frac{\sqrt{3}}{2} T_2$$

$$\Rightarrow T_2 = \sqrt{\frac{2}{3}} T_1 \tag{6.15}$$

Newton's second law on the y-axis:

$$\sum F_y = ma_y$$

$$\Rightarrow T_1 \sin 45° + T_2 \sin 30° - mg = ma$$

$$\Rightarrow \frac{\sqrt{2}}{2} T_1 + \frac{1}{2} T_2 = m(g + a) \tag{6.16}$$

Solving (6.15) and (6.16):

$$\Rightarrow \frac{\sqrt{2}}{2} T_1 + \frac{1}{2} \sqrt{\frac{2}{3}} T_1 = m(g + a)$$

$$\Rightarrow \frac{\sqrt{2}}{2} T_1 \left(1 + \frac{1}{\sqrt{3}} \right) = m(g + a)$$

$$\Rightarrow \frac{\sqrt{2}}{2} T_1 \left(\frac{\sqrt{3} + 1}{\sqrt{3}} \right) = m(g + a)$$

$$\Rightarrow T_1 = \frac{2}{\sqrt{2}} \left(\frac{\sqrt{3}}{\sqrt{3} + 1} \right) m(g + a)$$

$$\Rightarrow T_1 = 2 \sqrt{\frac{3}{2}} \left(\frac{1}{\sqrt{3} + 1} \right) m(g + a)$$

$$\Rightarrow T_1 = 2 \sqrt{\frac{3}{2}} \left(\frac{1}{\sqrt{3} + 1} \times \frac{\sqrt{3} - 1}{\sqrt{3} - 1} \right) m(g + a)$$

$$\Rightarrow T_1 = \sqrt{\frac{3}{2}} \left(\sqrt{3} - 1 \right) m(g + a)$$

Choice (4) is the answer.

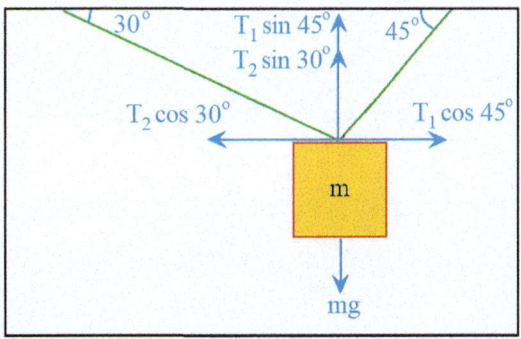

Fig. 6.12 The figure concerned with Problem 6.12

Notes

In this problem, the relations below have been used.

$$\cos 45^\circ = \frac{\sqrt{2}}{2}$$

$$\cos 30^\circ = \frac{\sqrt{3}}{2}$$

$$\sin 45^\circ = \frac{\sqrt{2}}{2}$$

$$\sin 30^\circ = \frac{1}{2}$$

6.4 Centripetal Force

6.13. Figure 6.13 shows all the forces applied on the object. To not fall from the wall, the static friction must be larger than the weight of the car. In other words:

$$f_s > mg \Rightarrow \mu_s N > mg \tag{6.17}$$

Also, by writing Newton's second law on the radial axis, we have:

$$\sum F = ma$$

$$\Rightarrow N - 0 = m\frac{v^2}{R} \Rightarrow N = m\frac{v^2}{R} \tag{6.18}$$

Solving (6.17) and (6.18):

$$\mu_s m \frac{v^2}{R} > mg \Rightarrow v^2 \geq \frac{Rg}{\mu} \Rightarrow v \geq \sqrt{\frac{Rg}{\mu}}$$

$$\Rightarrow v_{min} = \sqrt{\frac{Rg}{\mu}}$$

Choice (4) is the answer.

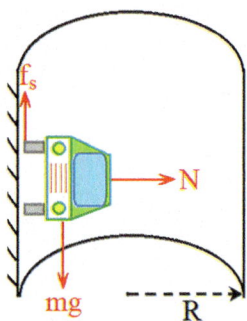

Fig. 6.13 The figure concerned with Problem 6.13

Notes

In this problem, the relation below has been used.

The centripetal acceleration, presented below, points towards the center of the circle in a circular motion. Herein, v and R are the linear velocity and radius of the circle, respectively.

$$a = \frac{v^2}{R}$$

6.5   Momentum

6.14. Based on the information given in the problem, we have (6.14):

$$m = 60 \ kg$$

$$t = 30 \ s$$

$$v_1 = 0 \ m/s$$

The area under the force-time curve of an object is its momentum change.

$$Area = p_2 - p_1$$

$$\Rightarrow Area = mv_2 - mv_1 = m(v_2 - v_1)$$

$$\Rightarrow \frac{1}{2} \times 40 \times 30 = 60(v_2 - 0) \Rightarrow 600 = 60v_2$$

$$\Rightarrow v_2 = 10 \ m/s$$

Choice (4) is the answer.

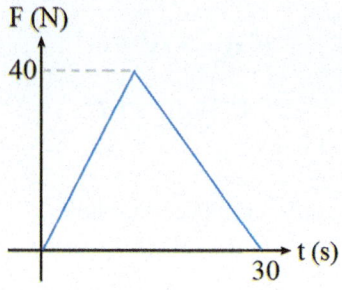

Fig. 6.14 The figure concerned with Problem 6.14

6.6   Position-Dependent Force and Potential Energy

6.15. The amount of work done on an object under the force $F(x)$, to move it from x_1 to x_2, can be calculated as follows.

$$W = \int_{x_1}^{x_2} F(x)dx$$

$$\Rightarrow W = \int_{1}^{2} \left(-x + \frac{1}{x^3} \right) dx = -\frac{x^2}{2} - \frac{1}{2x^2} \Big]_{1}^{2}$$

$$\Rightarrow W = \left(\frac{1^2}{2} + \frac{1}{2 \times 1^2} \right) - \left(\frac{2^2}{2} + \frac{1}{2 \times 2^2} \right)$$

$$\Rightarrow W = \left(\frac{1}{2} + \frac{1}{2} \right) - \left(2 + \frac{1}{8} \right)$$

$$\Rightarrow W = -\frac{9}{8} \ J$$

Choice (2) is the answer.

> **Notes**
>
> In this problem, the relations below have been used.
>
> $$\int x^n dx = \frac{1}{n+1}x^{n+1} + c$$

6.16. Based on the information given in the problem, we have:

$$F(x) = -3x^2 - 8x^3$$

$$U(x=0) = 1 \, J$$

As we know, the potential energy of a spring can be calculated as follows.

$$U(x) = -\int F dx$$

Therefore:

$$U(x) = -\int \left(-3x^2 - 8x^3\right)dx = x^3 + 2x^4 + c$$

By applying the boundary condition, we have:

$$U(x=0) = (0)^3 + 2(0)^4 + c = 1 \Rightarrow c = 1$$

$$\Rightarrow U(x) = x^3 + 2x^4 + 1$$

Hence, for $x = 1$, we have:

$$U(x=1) = (1)^3 + 2(1)^4 + 1$$

$$\Rightarrow U(x=1) = 4 \, J$$

Choice (3) is the answer.

> **Notes**
>
> In this problem, the relation below has been used.
>
> $$\int x^n dx = \frac{1}{n+1}x^{n+1} + c$$

6.7   Conservative Force

6.17. A force vector $\vec{F}$ is conservative if its curl is zero. In other words:

$$\nabla \times \vec{F} = 0$$

Therefore:

$$\begin{vmatrix} \widehat{i} & \widehat{j} & \widehat{k} \\ \dfrac{\partial}{\partial x} & \dfrac{\partial}{\partial y} & \dfrac{\partial}{\partial z} \\ ax + 2by^2 & cxy & 0 \end{vmatrix} = 0$$

$$\Rightarrow \left(0 - \frac{\partial}{\partial z}(cxy)\right)\widehat{i} + \left(\frac{\partial}{\partial z}(ax + 2by^2) - 0\right)\widehat{j} + \left(\frac{\partial}{\partial x}(cxy) - \frac{\partial}{\partial y}(ax + 2by^2)\right)\widehat{k} = 0$$

$$\Rightarrow (0 - 0)\widehat{i} + (0 - 0)\widehat{j} + (cy - 4by)\widehat{k} = 0$$

$$\Rightarrow cy - 4by = 0 \Rightarrow c = 4b$$

Choice (4) is the answer.

> **Notes**
>
> In this problem, the relation below has been used.
>
> $$\frac{d}{dx}x^n = nx^{n-1}$$

6.8   Work-Energy Theorem

6.18. Based on the information given in the problem, we have:

$$\vec{r} = (3t + 5t^3)\widehat{i} \; m/s$$

$$0 \leq t \leq 1 \; s$$

$$m = 2 \; kg$$

The work-energy theorem states that the net work done by the forces on an object equals the change in its kinetic energy. In other words:

$$W = \Delta K \Rightarrow W = \frac{1}{2}m\left(v^2 - v_0^2\right)$$

$$\Rightarrow W = \frac{1}{2} \times 2 \times \left([v(t=1)]^2 - [v(t=0)]^2\right)$$

As we know, velocity is the derivative of position of an object. Therefore:

$$\vec{v}(t) = \frac{d\vec{r}(t)}{dt} = \left(3 + 15t^2\right)\hat{i}$$

Hence:

$$W = (3 + 15)^2 - (3 + 0)^2 = 18^2 - 3^2$$

$$\Rightarrow W = 315\,J$$

Choice (1) is the answer.

Notes

In this problem, the relations below has been used.

$$\frac{d}{dx}x^n = nx^{n-1}$$

6.9   Conservation of Mechanical Energy Principle

6.19. Based on the information given in the problem, we have:

$$h_1 = 0.5\,m$$

$$h_2 = 2\,m$$

$$g = 10\,m/s^2$$

The maximum velocity of the chair will happen at its minimum height. Based on the principle of conservation of mechanical energy, we can write:

$$E_1 = E_2$$

$$mgh_1 + \frac{1}{2}mv_{max}^2 = mgh_2 + 0$$

$$\Rightarrow gh_1 + \frac{1}{2}v_{max}^2 = gh_2$$

$$\Rightarrow 10 \times 0.5 + 0.5 \times v_{max}^2 = 10 \times 2$$

$$\Rightarrow v_{max}^2 = 30 \Rightarrow v_{max} \simeq 5.4 \; m/s$$

Choice (1) is the answer.

6.10 Spring Force and Hooke's Law

6.20. Based on the information given in the problem, we have:

$$k = 6 \; N/m$$

$$f = 3 \; Hz$$

If a spring is halved, the stiffness constant of each part will double. Hence:

$$k_1 = k_2 = 12 \; N/m$$

Moreover, if two springs are connected in parallel (shown in Fig. 6.15), their stiffness constants are added. Therefore:

$$k_{tot} = k_1 + k_2 = 12 + 12 = 24 \; N/m$$

On the other hand, the angular frequency of the oscillations of a spring can be calculated as follows.

$$\omega = \sqrt{\frac{k}{m}}$$

Thus, for this problem we can write:

$$2\pi f = \sqrt{\frac{k_{tot}}{m}}$$

$$\Rightarrow 2\pi \times 3 = \sqrt{\frac{24}{m}}$$

$$\Rightarrow 36\pi^2 = \frac{24}{m} \Rightarrow m = \frac{24}{36\pi^2}$$

$$\Rightarrow m = \frac{2}{3\pi^2} \ kg$$

Choice (4) is the answer.

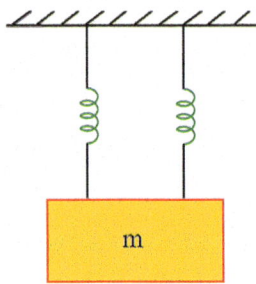

Fig. 6.15 The figure concerned with Problem 6.20

Notes

In this problem, the relation below has been used.

$$\omega = 2\pi f$$

■ ■ ■

References

1. Rahmani-Andebili, M. (2023). Calculus III – Practice Problems, Methods, and Solutions, Springer Nature.
2. Rahmani-Andebili, M. (2023). Calculus II – Practice Problems, Methods, and Solutions, Springer Nature.
3. Rahmani-Andebili, M. (2023). Calculus I (2nd Ed.) – Practice Problems, Methods, and Solutions, Springer Nature.
4. Rahmani-Andebili, M. (2024). Precalculus (2nd Ed.) – Practice Problems, Methods, and Solutions, Springer Nature.

Abstract

In this chapter, the basic and advanced problems of collision, center of mass, and theorem of Pappus are studied. Herein, different types of problems and exercises are presented that are categorized as follows.

- *Problems with detailed solution*: They have been designed to teach students the subjects in detail. Moreover, they have been categorized in different levels based on their difficulty levels (easy, normal, and hard) and calculation amounts (small, normal, and large).
- *Partially solved exercises*: They have been designed to encourage students to practice problems while guiding them through the problem-solving procedure and hinting the required formulas.
- *Exercises with final answer*: They have been designed to encourage students to practice more by themselves while hinting them by the final answer as well as to help instructors to give tests or quizzes.

7.1   Collision

Problem

7.1. A 2 *kg* ball is freed from the height of 0.4 *m*, and then hits a spring with the 1960 *N/m* stiffness constant. Calculate the maximum compression of the spring [1–4]. Herein, assume that $g = 9.8$ *m/s*^{2} (Fig. 7.1).

Difficulty level   ○ Easy   ○ Normal   ● Hard
Calculation amount   ○ Small   ● Normal   ○ Large

1) 8.9 *cm*
2) 10 *cm*
3) 0.12 *m*
4) 0.2 *m*

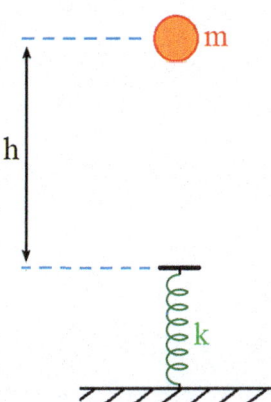

Fig. 7.1 The figure concerned with Problem 7.1

Exercise

A bullet with the mass of m hits a pendulum with the mass of M, penetrates it, and remains there. Calculate the initial velocity of the bullet if the pendulum rises to the heigh of h.

Final Answer

$$v_0 = \frac{m + M}{m}\sqrt{2gh}$$

Problem

7.2. In Fig. 7.2, that shows two balls with the masses of $m_A = 5\ kg$ and $m_B = 3\ kg$, assume that ball A with the velocity 2 m/s hits the other one, which is in stationary state. After the collision, ball A is deviated with the angle 30° (with respect to the positive direction of x-axis) and moves with 1 m/s velocity. Calculate the velocity and the deviation angle of ball B with respect to the positive direction of x-axis.

Difficulty level   ○ Easy   ○ Normal   ● Hard
Calculation amount   ○ Small   ○ Normal   ● Large
1) 2.1 m/s, 0°
2) 1.89 m/s, 24°
3) 0.83 m/s, − 24°
4) 2.1 m/s, − 24°

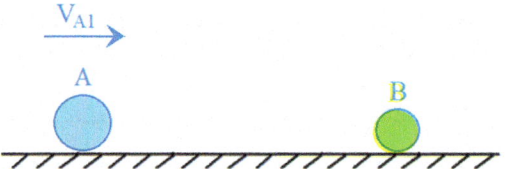

Fig. 7.2 The figure concerned with Problem 7.2

Problem

7.3. Figure 7.3 shows two objects ($m_A = 0.5\ kg$ and $m_B = 0.3\ kg$) that move toward each other on a frictionless surface with the initial velocities $v_{A1} = 2\ m/s$ and $v_{B1} = 2\ m/s$. If the objects stick to one another after the collision, calculate their final speed and the change in their total kinetic energy.

Difficulty level   ○ Easy   ○ Normal   ● Hard
Calculation amount   ○ Small   ○ Normal   ● Large
1) 0.5 m/s, − 1.5 J
2) 0 m/s, − 1.6 J
3) 1 m/s, 1.6 J
4) 0 m/s, − 1.5 J

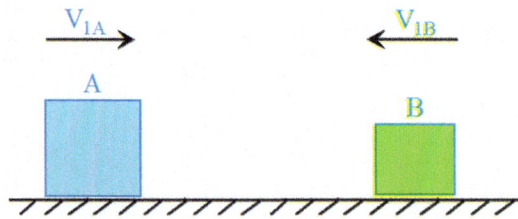

Fig. 7.3 The figure concerned with Problem 7.3

Partially Solved Exercise

A bullet with the mass m and velocity v hits a wall and come back with its half velocity. Calculate the magnitude of change in its momentum.

Solution

Based on the information given in the problem, we have:

$$v_1 = v$$

$$v_2 = -\frac{v}{2}$$

Therefore:

$$\Delta p = mv_2 - mv_1 = m(v_2 - v_1) = m((\quad) - (\quad)) = (\quad)$$

$$\Rightarrow |\Delta p| = \frac{3}{2}mv$$

Problem

7.4. A bullet with the mass of m and initial velocity of v_0 hits an object with the mass of $M = 2m$, and after penetration, exits from it with the velocity of $0.5v_0$. Calculate the amount of heat created by this experiment.

Difficulty level ○ Easy ○ Normal ● Hard
Calculation amount ○ Small ○ Normal ● Large

1) $\frac{5}{16}mv_0^2$

2) $\frac{7}{16}mv_0^2$

3) $\frac{3}{4}mv_0^2$

4) $\frac{5}{8}mv_0^2$

Problem

7.5. A pendulum with the mass of m_1 is freed from the height of h. The pendulum hits a stationary object with the mass of $m_2 = 2m_1$ at its lowest point. Calculate the height that the pendulum rises if the collision is perfectly inelastic.

Difficulty level ○ Easy ○ Normal ● Hard
Calculation amount ○ Small ○ Normal ● Large

1) $\frac{1}{3}h$

2) $\frac{1}{4}h$

3) $\frac{1}{9}h$

4) $\frac{4}{5}h$

Partially Solved Exercise

A bullet horizontally hits an object located on a frictionless surface. If the bullet sticks to the object and 40% of its initial energy is wasted, calculate the ratio of mass of the object to the one of the bullet.

Solution

Since the bullet sticks to the object, the collision is perfectly inelastic. Therefore, the total momentum of the system is conserved while the total kinetic energy of the system is not conserved.

By applying the conservation of mechanical energy principle for the system and considering that 40% of the initial energy of the bullet is wasted, we have:

$$0.6 \times E_1 = E_2$$

$$\Rightarrow 0.6 \times \frac{1}{2} (\qquad) v^2 = \frac{1}{2} (\qquad) v'^2$$

$$\Rightarrow 0.6 m v^2 = (m + M) v'^2 \tag{7.1}$$

By applying the conservation of momentum principle for the system, we have:

$$p_1 = p_2$$

$$\Rightarrow (\qquad) v = (\qquad) v'$$

$$\Rightarrow v = \frac{m + M}{m} v' \tag{7.2}$$

Solving (7.1) and (7.2):

$$\Rightarrow \qquad\qquad =$$

$$\Rightarrow \qquad\qquad =$$

$$\Rightarrow \qquad\qquad =$$

$$\Rightarrow \frac{M}{m} = \frac{2}{3}$$

7.2   Centre of Mass and Theorem of Pappus

Problem

7.6. Calculate the center of mass of a semicircular disk with the radius a located at the origin, shown in Fig. 7.4.

Difficulty level   ○ Easy   ○ Normal   ● Hard
Calculation amount   ● Small   ○ Normal   ○ Large

1) $x_{cm} = 0, y_{cm} = \dfrac{4a}{3\pi}$

2) $x_{cm} = 0, y_{cm} = \dfrac{2a}{3\pi}$

3) $x_{cm} = 0, y_{cm} = \dfrac{a}{3\pi}$

4) $x_{cm} = 0, y_{cm} = \dfrac{a}{\pi}$

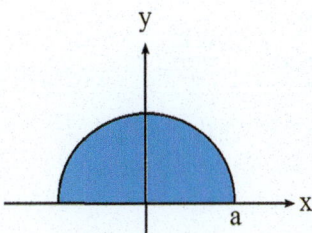

Fig. 7.4   The figure concerned with Problem 7.6

Problem

7.7. Figure 7.5 illustrates a thin rod with the length and linear density of L and $\rho = \dfrac{\rho_0}{L^2}x^2$, respectively. Calculate its center of mass.

Difficulty level   ○ Easy   ○ Normal   ● Hard

Calculation amount   ○ Small   ● Normal   ○ Large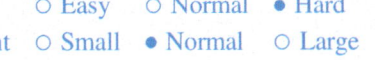

1) $x_{cm} = \dfrac{5L}{6}$

2) $x_{cm} = \dfrac{3L}{4}$

3) $x_{cm} = \dfrac{2L}{3}$

4) $x_{cm} = \dfrac{L}{2}$

Fig. 7.5   The figure concerned with Problem 7.7

Exercise

Calculate the center of mass of a thin rod with the length and linear density of L and $\rho = x^3$, respectively.

Final Answer

$x_{cm} = \dfrac{4L}{5}$

Problem

7.8. Calculate the center of mass of a rod with the linear density λ that has been bent to make a semicircle with the radius r at the origin. See Fig. 7.6.

Difficulty level ○ Easy ○ Normal ● Hard
Calculation amount ○ Small ○ Normal ● Large

1) $x_{cm} = 0, y_{cm} = \dfrac{r}{\pi}$

2) $x_{cm} = \dfrac{2r}{\pi}, y_{cm} = \dfrac{2r}{\pi}$

3) $x_{cm} = 0, y_{cm} = \dfrac{2r}{\pi}$

4) $x_{cm} = 0, y_{cm} = 0$

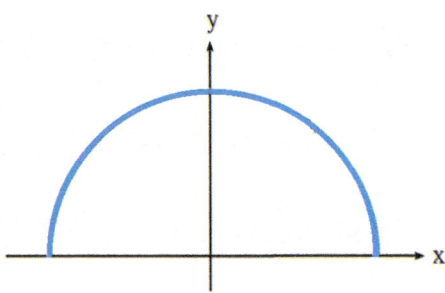

Fig. 7.6 The figure concerned with Problem 7.8

Exercise

Calculate the center of mass of a rod with the linear density λ that has been bent to make a quarter circle (quadrant) with the radius r at the origin.

Final Answer

$(x_{cm}, y_{cm}) = \left(\dfrac{2r}{\pi}, \dfrac{2r}{\pi} \right)$

Problem

7.9. Figure 7.7 shows a rectangle surface with the surface density $\sigma(x, y) = \sigma_0 x^2 y$. Calculate its center of mass.

Difficulty level ○ Easy ○ Normal ● Hard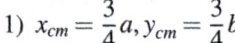
Calculation amount ○ Small ○ Normal ● Large

1) $x_{cm} = \dfrac{3}{4}a, y_{cm} = \dfrac{3}{4}b$

2) $x_{cm} = \dfrac{3}{4}a, y_{cm} = \dfrac{2}{3}b$

3) $x_{cm} = \dfrac{2}{3}a, y_{cm} = \dfrac{2}{3}b$

4) $x_{cm} = \dfrac{2}{3}a, y_{cm} = \dfrac{3}{4}b$

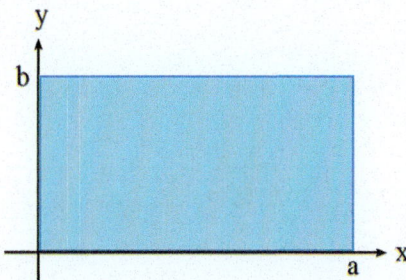

Fig. 7.7 The figure concerned with Problem 7.9

Exercise

Calculate its center of mass of a rectangle surface with the surface density $\sigma(x, y) = xy$.

Final Answer

$$(x_{cm}, y_{cm}) = \left(\frac{2}{3}a, \frac{2}{3}b\right)$$

Exercise

Calculate the center of mass of a solid semi-sphere with radius R and total mass of M.

Final Answer

$$(x_{cm}, y_{cm}, z_{cm}) = \left(0, \frac{3}{8}R, 0\right)$$

Notes

In this problem, the relations below are needed.

The center of mass of a three-dimensional continuous distribution of mass can be calculated as follows.

$$x_{cm} = \frac{\iiint x\,dm}{\iiint dm}$$

$$y_{cm} = \frac{\iiint y\,dm}{\iiint dm}$$

$$z_{cm} = \frac{\iiint z\,dm}{\iiint dm}$$

References

1. Rahmani-Andebili, M. (2023). Calculus III – Practice Problems, Methods, and Solutions, Springer Nature.
2. Rahmani-Andebili, M. (2023). Calculus II – Practice Problems, Methods, and Solutions, Springer Nature.
3. Rahmani-Andebili, M. (2023). Calculus I (2nd Ed.) – Practice Problems, Methods, and Solutions, Springer Nature.
4. Rahmani-Andebili, M. (2024). Precalculus (2nd Ed.) – Practice Problems, Methods, and Solutions, Springer Nature.

Collision and Centre of Mass: Part B

8

Abstract

In this chapter, the problems of the seventh chapter are fully solved, in detail, step-by-step, and with different methods.

8.1 Collision

8.1. Based on the information given in the problem, we have:

$$m = 2 \; kg$$

$$h = 0.4 \; m$$

$$k = 1960 \; N/m$$

$$g = 9.8 \; m/s^2$$

The maximum compression of the spring happens when the whole gravitational potential energy of the ball is converted to the elastic potential energy of the spring [1–4].

By applying the conservation of mechanical energy principal, we have:

$$E_1 = E_2$$

$$\Rightarrow mg(h + x) = \frac{1}{2}kx^2$$

$$\Rightarrow 2 \times 9.8(0.4 + x) = \frac{1}{2} \times 1960x^2$$

$$\Rightarrow 0.4 + x = 50x^2$$

$$\Rightarrow 50x^2 - x - 0.4 = 0$$

$$\Rightarrow x = \frac{1 \pm \sqrt{1+80}}{100} = \frac{1 \pm 9}{100} = 0.1, \ -0.08 \ m$$

Only $x = 0.1 \ m$ is acceptable which is $x = 10 \ cm$.
Choice (2) is the answer.

Note that if the compression of spring (x) in the gravitational potential energy of the ball was ignored, the misguiding answer of 8.9 cm would be achieved (Fig. 8.1).

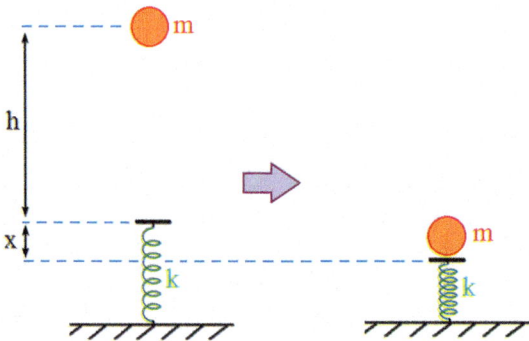

Fig. 8.1 The figure concerned with Problem 8.1

Notes

In this problem, the relations below have been used.

$$ax^2 + bx + c = 0 \Rightarrow x = \frac{-b \pm \sqrt{b^2 - 4ac}}{2a}$$

The elastic potential energy of a spring can be calculated as follows, in which x is amount of compression or stretch of the spring in SI system.

$$U(x) = \frac{1}{2}kx^2$$

8.2. Based on the information given in the problem, we have:

$$m_A = 5 \ kg$$

$$m_B = 3 \ kg$$

$$v_{A1} = 2 \ m/s$$

$$\begin{cases} v_{A1x} = 2 \ m/s \\ v_{A1y} = 0 \ m/s \end{cases}$$

$$\begin{cases} v_{B1x} = 0 \ m/s \\ v_{B1y} = 0 \ m/s \end{cases}$$

$$\begin{cases} v_{A2x} = 1 \times \cos 30\,° = \dfrac{\sqrt{3}}{2} \ m/s \\[3mm] v_{A2y} = 1 \times \sin 30\,° = \dfrac{1}{2} \ m/s \end{cases}$$

Since there is no external force, the total momentum of the system will be conserved on both axes.

Total momentum of the system on x-axis:

$$p_{1x} = p_{2x}$$

$$\Rightarrow m_A v_{A1x} + m_B v_{B1x} = m_A v_{A2x} + m_B v_{B2x}$$

$$\Rightarrow 5 \times 2 + 3 \times 0 = 5 \times \frac{\sqrt{3}}{2} + 3 v_{B2x}$$

$$\Rightarrow v_{B2x} = 1.89 \ m/s$$

Total momentum of the system on y-axis:

$$p_{1y} = p_{2y}$$

$$\Rightarrow m_A v_{A1y} + m_B v_{B1y} = m_A v_{A2y} + m_B v_{B2y}$$

$$\Rightarrow 5 \times 0 + 3 \times 0 = 5 \times \frac{1}{2} + 3 v_{B2y}$$

$$\Rightarrow v_{B2x} = -0.83 \ m/s$$

$$\Rightarrow v_{B2} = \sqrt{v_{B2x}^2 + v_{B2y}^2} = \sqrt{1.89^2 + (-0.83)^2}$$

$$\Rightarrow v_{B2} = 2.1 \ m/s$$

$$\beta = \arctan \frac{v_{B2y}}{v_{B2x}} = -\arctan \frac{0.83}{1.89}$$

$$\Rightarrow \beta = -24\,°$$

Choice (4) is the answer (Fig. 8.2).

Fig. 8.2 The figure concerned with Problem 8.2

8.3. Based on the information given in the problem, we have:

$$m_A = 0.5 \; kg$$

$$m_B = 0.3 \; kg$$

$$v_{A1} = 2 \; m/s$$

$$v_{B1} = -2 \; m/s$$

Since the objects stick to each other, the collision is perfectly inelastic. Therefore, the total momentum of the system will be conserved; however, the total kinetic energy of the system will not be conserved.

$$p_1 = p_2$$

$$\Rightarrow m_A v_{1A} + m_B v_{1B} = (m_A + m_B)v_2$$

$$\Rightarrow v_2 = \frac{m_A v_{1A} + m_B v_{1B}}{m_A + m_B}$$

$$\Rightarrow v_2 = \frac{(0.5)(2) + (0.3)(-2)}{0.5 + 0.3} \Rightarrow v_2 = 0.5 \; m/s$$

Regarding the kinetic energy of the system, we have:

$$K_{1A} = \frac{1}{2} m_A v_{1A}^2 = \frac{1}{2}(0.5)(2)^2 = 1\ J$$

$$K_{1B} = \frac{1}{2} m_B v_{1B}^2 = \frac{1}{2}(0.3)(-2)^2 = 0.6\ J$$

$$\Rightarrow K_1 = K_{1A} + K_{1B} = 1.6\ J$$

$$K_2 = \frac{1}{2}(m_A + m_B)v_2^2 = \frac{1}{2}(0.5 + 0.3)(0.5)^2 = 0.1\ J$$

$$\Delta K = K_2 - K_1 = 0.1 - 1.6$$

$$\Rightarrow \Delta K = -1.5\ J$$

Choice (1) is the answer (Fig. 8.3).

Fig. 8.3 The figure concerned with Problem 8.3

Notes

In this problem, the relation below has been used.

A perfectly inelastic collision is a collision in which the two objects stick to one another. In this regard, the total momentum of the system is conserved; however, the total kinetic energy of the system is not conserved.

$$K = \frac{1}{2} m v^2$$

8.4. Based on the conservation of momentum, we have:

$$p_1 = p_2$$

$$\Rightarrow m v_0 = M v + m \frac{v_0}{2}$$

$$\Rightarrow m v_0 = (2m)v + m \frac{v_0}{2} \Rightarrow m \frac{v_0}{2} = 2mv$$

$$\Rightarrow v = \frac{v_0}{4} \tag{8.1}$$

Based on the conservation of mechanical energy, we have:

$$E_1 = E_2 + Q$$

$$\Rightarrow \frac{1}{2}mv_0^2 = \frac{1}{2}Mv^2 + \frac{1}{2}m\left(\frac{v_0}{2}\right)^2 + Q$$

$$\Rightarrow Q = \frac{1}{2}mv_0^2 - \frac{1}{2}(2m)v^2 - \frac{1}{8}mv_0^2$$

$$\Rightarrow Q = \frac{3}{8}mv_0^2 - mv^2 \tag{8.2}$$

Solving (8.1) and (8.2):

$$Q = \frac{3}{8}mv_0^2 - m\left(\frac{v_0}{4}\right)^2 = \frac{3}{8}mv_0^2 - \frac{1}{16}mv_0^2$$

$$\Rightarrow Q = \frac{5}{16}mv_0^2$$

Choice (1) is the answer.

8.5. Based on the information given in the problem, we know that the collision is perfectly inelastic. Therefore, the object sticks to the pendulum and total momentum of the system is conserved.

By applying the conservation of mechanical energy principle for the pendulum, we can calculate the velocity of pendulum before hitting the object as follows.

$$E_1 = E_2$$

$$m_1 gh = \frac{1}{2}m_1 v_1^2 \Rightarrow v_1 = \sqrt{2gh}$$

By applying the conservation of momentum principle for the combination of pendulum and object we can calculate their velocity right after the collision as follows.

$$p_1 = p_2$$

$$m_1 v_1 + 0 = (m_1 + m_2)v_2$$

$$\Rightarrow m_1 \sqrt{2gh} = (m_1 + 2m_1)v_2$$

$$\Rightarrow v_2 = \frac{1}{3}\sqrt{2gh}$$

By applying the conservation of mechanical energy for the combination of pendulum and object we can calculate their height as follows.

$$E_2 = E_3$$

$$\Rightarrow \frac{1}{2}(m_1 + m_2)v_2^2 = (m_1 + m_2)gh'$$

$$\Rightarrow \frac{1}{2}v_2^2 = gh'$$

$$\Rightarrow \frac{1}{2}\left(\frac{1}{3}\sqrt{2gh}\right)^2 = gh'$$

$$\Rightarrow h' = \frac{1}{9}h$$

Choice (3) is the answer.

8.2   Centre of Mass and Theorem of Pappus

8.6. Theorem of Pappus for volume states that if A and l, respectively, are a region and a line in the plane that does not intersect each other, as is illustrated in Fig. 8.4a, the volume of the solid of revolution formed by revolving A around l is equal to the area of A multiplied by the distance traveled by the center of mass of A. In other words:

$$V = 2\pi y_{cm}A$$

Now, the center of mass of the solid can be calculated as follows.

$$y_{cm} = \frac{V}{2\pi A}$$

By applying Theorem of Pappus for volume in this problem, shown in Fig. 8.4b, the semicircular disk needs to revolve around its dimeter that results in a solid sphere with the same radius of the disk. Hence:

$$y_{cm} = \frac{\frac{4}{3}\pi a^3}{2\pi \frac{\pi a^2}{2}} \Rightarrow y_{cm} = \frac{4a}{3\pi}$$

On the other hand, $x_{cm} = 0$ due to the symmetry around the y-axis. Choice (1) is the answer.

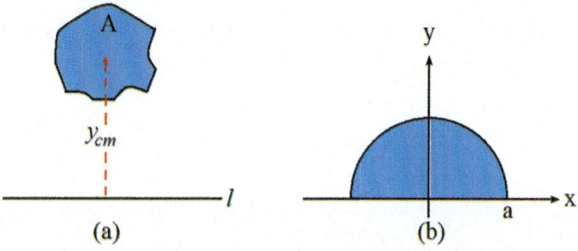

Fig. 8.4 The figure concerned with Problem 8.6

8.7. As we know the center of mass of a one-dimensional continuous distribution of mass can be calculated as follows.

$$x_{cm} = \frac{\displaystyle\int x\,dm}{\displaystyle\int dm}$$

Hence:

$$x_{cm} = \frac{\displaystyle\int_0^L x\rho\,dx}{\displaystyle\int_0^L \rho\,dx} = \frac{\displaystyle\int_0^L x\frac{\rho_0}{L^2}x^2\,dx}{\displaystyle\int_0^L \frac{\rho_0}{L^2}x^2\,dx} = \frac{\displaystyle\int_0^L x^3\,dx}{\displaystyle\int_0^L x^2\,dx} = \frac{\dfrac{L^4}{4}}{\dfrac{L^3}{3}}$$

$$\Rightarrow x_{cm} = \frac{3L}{4}$$

Choice (2) is the answer (Fig. 8.5).

Fig. 8.5 The figure concerned with Problem 8.7

8.8. Based on the information given in the problem and Fig. 8.6, we can write:

$$\begin{cases} x = r\cos\theta \\ y = r\sin\theta \end{cases}$$

$$0 \leq \theta \leq \pi$$

$$\begin{cases} dm = \lambda dl \\ dl = rd\theta \end{cases} \Rightarrow dm = \lambda rd\theta$$

As we know, the center of mass of a one-dimensional continuous distribution of mass can be calculated as follows.

$$x_{cm} = \frac{\displaystyle\int xdm}{\displaystyle\int dm}$$

$$y_{cm} = \frac{\displaystyle\int ydm}{\displaystyle\int dm}$$

Since the problem is related to a semicircle, it is better to solve it in polar coordinates system.

$$x_{cm} = \frac{\displaystyle\int r\cos\theta \times \lambda rd\theta}{\displaystyle\int \lambda rd\theta} = \frac{r\displaystyle\int_0^\pi \cos\theta d\theta}{\displaystyle\int_0^\pi d\theta}$$

$$\Rightarrow x_{cm} = \frac{r\sin\theta\big|_0^\pi}{\theta\big|_0^\pi} = \frac{r(\sin\pi - \sin 0)}{\pi - 0}$$

$$x_{cm} = 0$$

$x_{cm} = 0$ could be achieved without calculation because the semicircle is symmetric around the y-axis.

$$y_{cm} = \frac{\displaystyle\int r\sin\theta \times \lambda rd\theta}{\displaystyle\int \lambda rd\theta} = \frac{r\displaystyle\int_0^\pi \sin\theta d\theta}{\displaystyle\int_0^\pi d\theta}$$

$$\Rightarrow y_{cm} = \frac{r[-\cos\theta]_0^\pi}{\theta\big|_0^\pi} = \frac{r(-\cos\pi + \cos 0)}{\pi - 0}$$

$$\Rightarrow y_{cm} = \frac{2r}{\pi}$$

Choice (3) is the answer.

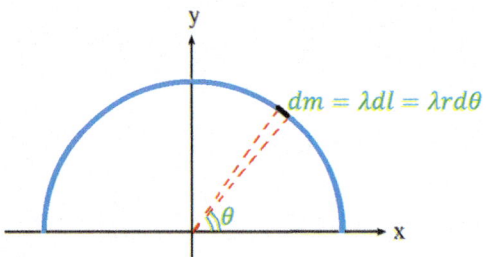

Fig. 8.6 The figure concerned with Problem 8.8

Notes

In this problem, the relations below have been used.

$$\int \cos\theta d\theta = \sin\theta$$

$$\int \sin\theta d\theta = -\cos\theta$$

$$\sin\pi = 0$$

$$\sin 0 = 0$$

$$\cos\pi = -1$$

$$\cos 0 = 1$$

8.9. Based on the information given in the problem and Fig. 8.7, we can write:

$$\sigma(x, y) = \sigma_0 x^2 y$$

$$\begin{cases} dm = \sigma ds \\ ds = dxdy \end{cases} \Rightarrow dm = \sigma dxdy$$

The center of mass of a two-dimensional continuous distribution of mass can be calculated as follows.

$$x_{cm} = \frac{\iint x dm}{\iint dm}$$

$$y_{cm} = \frac{\iint y dm}{\iint dm}$$

Since the problem is related to a rectangular surface, it is better to solve it in cartesian coordinates system.

$$x_{cm} = \frac{\iint x(\sigma_0 x^2 y)dxdy}{\iint \sigma_0 x^2 y dxdy} = \frac{\int_0^a x^3 dx \int_0^b y dy}{\int_0^a x^2 dx \int_0^b y dy} \Rightarrow x_{cm} = \frac{\frac{a^4}{4} \times \frac{b^2}{2}}{\frac{a^3}{3} \times \frac{b^2}{2}} \Rightarrow x_{cm} = \frac{3}{4}a$$

$$y_{cm} = \frac{\iint y(\sigma_0 x^2 y)dxdy}{\iint \sigma_0 x^2 y dxdy} = \frac{\int_0^a x^2 dx \int_0^b y^2 dy}{\int_0^a x^2 dx \int_0^b y dy} \Rightarrow y_{cm} = \frac{\frac{a^3}{3} \times \frac{b^3}{3}}{\frac{a^3}{3} \times \frac{b^2}{2}} \Rightarrow y_{cm} = \frac{2}{3}b$$

Choice (2) is the answer.

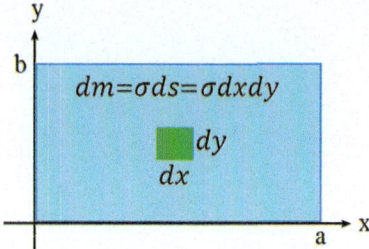

Fig. 8.7 The figure concerned with Problem 8.9

Notes

In this problem, the relation below has been used.

$$\int x^n dx = \frac{1}{n+1}x^{n+1} + c$$

References
1. Rahmani-Andebili, M. (2023). Calculus III – Practice Problems, Methods, and Solutions, Springer Nature.
2. Rahmani-Andebili, M. (2023). Calculus II – Practice Problems, Methods, and Solutions, Springer Nature.
3. Rahmani-Andebili, M. (2023). Calculus I (2nd Ed.) – Practice Problems, Methods, and Solutions, Springer Nature.
4. Rahmani-Andebili, M. (2024). Precalculus (2nd Ed.) – Practice Problems, Methods, and Solutions, Springer Nature.

Abstract

In this chapter, the basic and advanced problems of rotational kinematics and dynamics are studied. The subjects include rotational kinematics, moment of inertia, parallel axis theorem, rotational dynamics, rotational kinetic energy, conservation of mechanical energy principle, and work-energy theorem. Herein, different types of problems and exercises are presented that are categorized as follows.

- *Problems with detailed solution*: They have been designed to teach students the subjects in detail. Moreover, they have been categorized in different levels based on their difficulty levels (easy, normal, and hard) and calculation amounts (small, normal, and large).
- *Partially solved exercises*: They have been designed to encourage students to practice problems while guiding them through the problem-solving procedure and hinting the required formulas.
- *Exercises with final answer*: They have been designed to encourage students to practice more by themselves while hinting them by the final answer as well as to help instructors to give tests or quizzes.

9.1   Rotational Kinematics

Problem

9.1. Which one of the choices is correct about the system shown in Fig. 9.1. Herein, assume that $r_1 < r_2$ [1–4].

Difficulty level   ● Easy   ○ Normal   ○ Hard
Calculation amount   ● Small   ○ Normal   ○ Large

1) The angular velocity of both wheels is the same, but their linear velocity is different.
2) The liner velocity of both wheels is the same, but their angular velocity is different.
3) The angular and linear velocities of both wheels are the same.
4) The angular and linear velocities of both wheels are different.

Fig. 9.1 The figure concerned with Problem 9.1

Problem

9.2. The instantaneous position equation of an object with the mass of 2 kg is $\vec{r}(t) = 3t^2\hat{i} + 2t\hat{j}$. Calculate the resulted instantaneous torque.

Difficulty level   ○ Easy   ● Normal   ○ Hard
Calculation amount   ○ Small   ● Normal   ○ Large

1) $24t\hat{k}$
2) $12t\hat{k}$
3) $-24t\hat{k}$
4) $-12t\hat{k}$

Exercise

Calculate the torque resulted by a 1 kg body with $\vec{r}(t) = 0.5t^2\hat{i} + \hat{j}$ as its equation of motion.

Final Answer

$\vec{\tau} = -\hat{k}$

9.2   Moment of Inertia

Problem

9.3. Consider the solid system shown in Fig. 9.2 that includes three 5 kg masses and three 1 m weightless rods. Calculate the moment of inertia of the system around the axis.

Difficulty level   ● Easy   ○ Normal   ○ Hard
Calculation amount   ● Small   ○ Normal   ○ Large

1) $15\ kg.\,m^2$
2) $10\ kg.\,m^2$
3) $5\ kg.\,m^2$
4) $20\ kg.\,m^2$

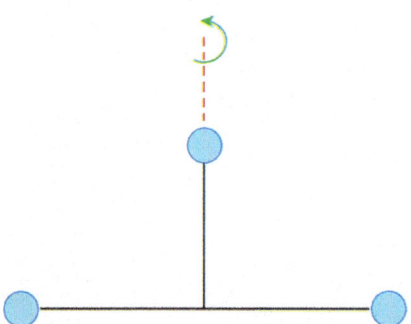

Fig. 9.2 The figure concerned with Problem 9.3

Partially Solved Exercise

Consider the solid system shown in Fig. 9.3 and calculate its moment of inertia around the requested axis.

Solution

The moment of inertia of the system can be calculated as follows.

$$I = \sum_{i=1}^{3} m_i r_i^2$$

$$\Rightarrow I = m_1 r_1^2 + m_2 r_2^2 + m_3 r_3^2$$

$$\Rightarrow I = (\quad) \times (\quad)^2 + (\quad) \times (\quad)^2 + (\quad) \times (\quad)^2$$

$$\Rightarrow I = 8 \ kg \cdot m^2$$

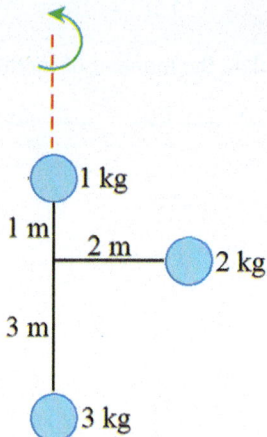

Fig. 9.3 The figure concerned with the partially solved exercise

Problem

9.4. Calculate the moment of inertia of a rod with the length and mass of L and M around the axis that perpendicularly passes from one of its ends (see Fig. 9.4).

Difficulty level   ○ Easy   ● Normal   ○ Hard
Calculation amount   ○ Small   ● Normal   ○ Large

1) $\frac{1}{2} ML^2$

2) $\frac{1}{3} ML^2$

3) $\frac{1}{12} ML^2$

4) $\frac{1}{24} ML^2$

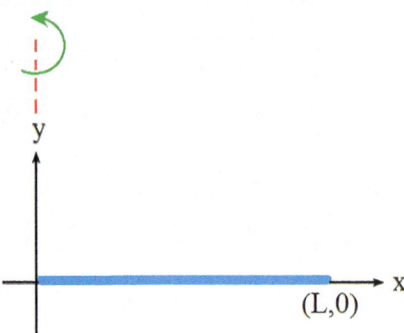

Fig. 9.4 The figure concerned with Problem 9.4

Partially Solved Exercise

In Problem 9.4, assume that $\lambda = x$, and then calculate the moment of inertia of the rod.

Solution

The moment of inertia of the rod can be calculated as follows.

$$I = \int r^2 \, dm$$

$$\Rightarrow I = \int_0^L x^2 \, dm$$

$$\Rightarrow I = \int_0^L x^2 (\quad)(\quad)$$

$$\Rightarrow I = \int_0^L (\quad) \, dx$$

$$\Rightarrow I = \left[\frac{(\quad)}{(\quad)} \right]_0^L$$

$$\Rightarrow I = \frac{L^4}{4}$$

Notes

In this problem, the relation below has been used.

$$\int x^n \, dx = \frac{1}{n+1} x^{n+1} + c$$

9.3   Parallel Axis Theorem

Problem

9.5. As can be seen in Fig. 9.5, a solid sphere with the mass and radius of 5 kg and 0.1 m has been connected to rod with the mass and length of 3 kg and 1 m. Calculate their total moment of inertia around the axis Δ.

Difficulty level   ○ Easy   ○ Normal   ● Hard

Calculation amount   ○ Small   ● Normal   ○ Large

1) 8.22 kgm^2
2) 6.02 kgm^2
3) 6.32 kgm^2
4) 7.07 kgm^2

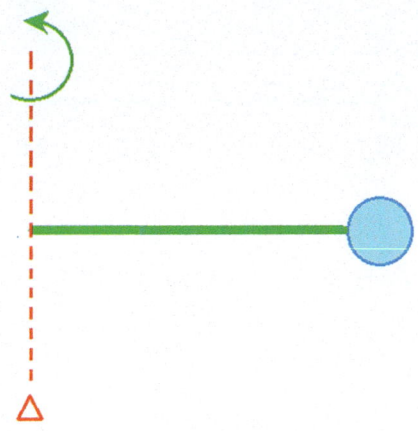

Fig. 9.5 The figure concerned with Problem 9.5

Partially Solved Exercise

Calculate the total moment of inertia of the system, illustrated in Fig. 9.6, around the axis Δ. Herein, the rods are assumed weightless.

Solution

As we know, the moment of inertia of a solid sphere (with the mass and radius of M and R) around its diameter can be calculated as follows.

$$I = \frac{2}{5}MR^2$$

As can be seen, the axis Δ and the axis of diameter of the solid spheres are different and far from each other. However, based on the parallel axis theorem the moment of inertia of each sphere can be calculated as follows.

$$I_1' = I_1 + M_1(L_1 + R_1)^2 = \frac{2}{5}M_1R_1^2 + M_1(L_1 + R_1)^2$$

$$I_2' = I_2 + M_2(L_2 + R_2)^2 = \frac{2}{5}M_2R_2^2 + M_2(L_2 + R_2)^2$$

Hence, the total moment of inertia of the system can be calculated as follows.

$$I_{total} = I'_1 + I'_2$$

$$\Rightarrow I_{total} = \frac{2}{5}M_1R_1^2 + M_1(L_1 + R_1)^2 + \frac{2}{5}M_2R_2^2 + M_2(L_2 + R_2)^2$$

$$\Rightarrow I_{total} = \frac{2}{5}(\quad)(\quad)^2 + (\quad)[(\quad) + (\quad)]^2 + \frac{2}{5}(\quad)(\quad)^2 + (\quad)[(\quad) + (\quad)]^2$$

$$\Rightarrow I_{total} = 10.926 \ kg.m^2$$

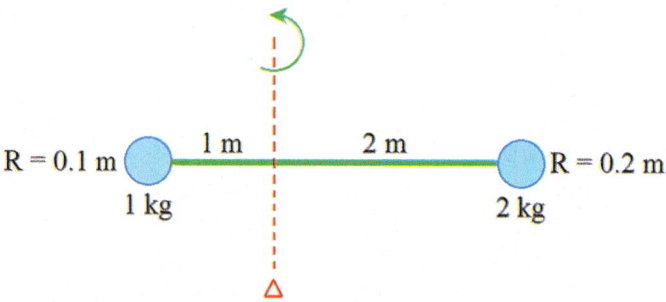

Fig. 9.6 The figure concerned with the partially solved exercise

9.4   Rotational Dynamics

Problem

9.6. In the Atwood machine shown in Fig. 9.7, $M = 2 \ kg$, $m_1 = 3 \ kg$, and $m_2 = 1 \ kg$. If the machine starts from the stationary state, the pulley is a solid cylinder, and the friction is negligible, calculate the acceleration of the masses. Herein, assume that $g = 10 \ m/s^2$ and the solid cylinder-shape pulley is not weightless.

Difficulty level   ○ Easy   ○ Normal   ● Hard
Calculation amount   ○ Small   ○ Normal   ● Large

1) $2 \ m/s^2$
2) $\frac{10}{3} \ m/s^2$
3) $4 \ m/s^2$
4) $5 \ m/s^2$

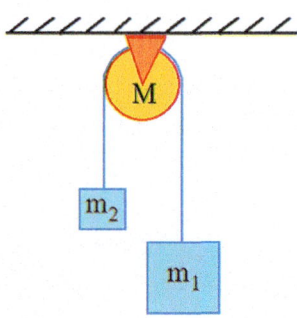

Fig. 9.7 The figure concerned with Problem 9.6

9.5 Rotational Kinetic Energy

9.7. A ring is rotating on a horizontal surface with a constant velocity. Calculate the ratio of its rotational kinetic energy to total kinetic energy (Fig. 9.8).

Difficulty level ○ Easy ● Normal ○ Hard
Calculation amount ○ Small ● Normal ○ Large

1) $\dfrac{1}{2}$

2) 2

3) $\dfrac{1}{3}$

4) 3

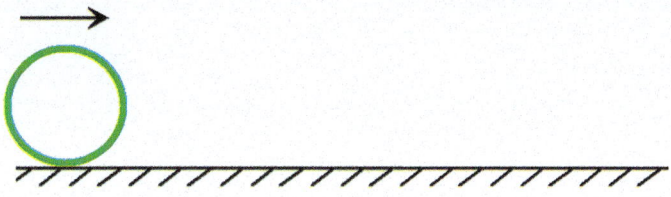

Fig. 9.8 The figure concerned with Problem 9.7

9.8. A solid sphere is rotating on a horizontal surface with a constant velocity. Calculate the ratio of its rotational kinetic energy to total kinetic energy (Fig. 9.9).

Difficulty level ○ Easy ● Normal ○ Hard
Calculation amount ○ Small ● Normal ○ Large

1) $\dfrac{2}{7}$

2) $\dfrac{2}{5}$

3) $\dfrac{5}{2}$

4) $\dfrac{7}{2}$

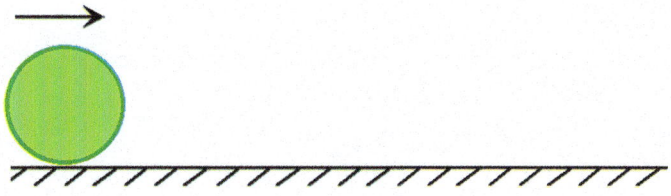

Fig. 9.9 The figure concerned with Problem 9.8

Partially Solved Exercise

A solid cylinder is rotating on a horizontal surface at a constant velocity. Calculate the ratio of its rotational kinetic energy to total kinetic energy.

Solution

As we know, the moment of inertia of a cylinder with the mass and radius of m and r around its axis is as follows.

$$I = \frac{1}{2}mr^2$$

In addition, the relation between the angular and linear velocities is as follows (Fig. 9.10).

$$v = r\omega$$

Therefore:

$$\frac{K_{rot}}{K_{tot}} = \frac{K_{rot}}{K_{lin} + K_{rot}} = \frac{\frac{1}{2}I\omega^2}{\frac{1}{2}mv_{cm}^2 + \frac{1}{2}I\omega^2}$$

$$\Rightarrow \frac{K_{rot}}{K_{tot}} = \frac{(\qquad\qquad)}{(\qquad\qquad) + (\qquad\qquad)} = \frac{(\quad)}{(\quad)}$$

$$\Rightarrow \frac{K_{rot}}{K_{tot}} = \frac{1}{3}$$

Fig. 9.10 The figure concerned with the partially solved exercise

9.6   Conservation of Mechanical Energy Principle

Problem

9.9. A 2 *kg* solid cylinder is freed from the top of a 4 *m* inclined surface with the angle of 30°. Calculate the velocity of the cylinder at the bottom of the inclined surface if its radius is 20 *cm* (Fig. 9.11).

Difficulty level   ○ Easy   ○ Normal   ● Hard

Calculation amount   ○ Small   ● Normal   ○ Large

1) $\sqrt{20}\ m/s$

2) $\sqrt{\frac{80}{3}}m/s$

3) $\frac{\sqrt{20}}{3}\ m/s$

4) $\sqrt{40}\ m/s$

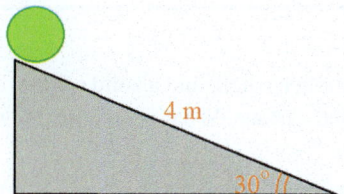

Fig. 9.11 The figure concerned with Problem 9.9

Exercise

In Problem 9.9, replace the solid cylinder with a solid sphere with the same radius and mass. Calculate the velocity of the sphere at the bottom of the inclined surface.

Final Answer

$$v_{cm} = 10\sqrt{\frac{2}{7}}\,m/s$$

Problem

9.10. A ring with the radius of R and mass of M is freed from the top of an inclined surface. Calculate its angular momentum with respect to the axis, passing from its center and perpendicular to its base, at the bottom of the inclined surface (Fig. 9.12).

Difficulty level   ○ Easy   ○ Normal   ● Hard
Calculation amount   ○ Small   ● Normal   ○ Large

1) $MR\sqrt{gh}$
2) $MR\sqrt{2gh}$
3) $MR\sqrt{\dfrac{gh}{3}}$
4) $2MR\sqrt{3gh}$

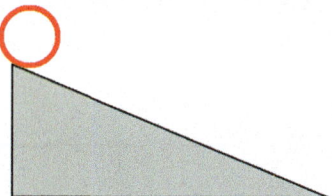

Fig. 9.12 The figure concerned with Problem 9.10

Exercise

In Problem 9.10, replace the ring with a circular disk with the same radius and mass. Calculate the angular momentum of the sphere at the bottom of the inclined surface.

Final Answer

$$L = \frac{2\sqrt{3}}{3}MR\sqrt{gh}$$

Problem

9.11. A cylindrical shell, a solid cylinder, a spherical shell, and a solid sphere, each with the mass and radius of M and R, are simultaneously freed from the top of an inclined surface. What is their arriving order?

Difficulty level   ○ Easy   ○ Normal   ● Hard
Calculation amount   ○ Small   ○ Normal   ● Large

1) The cylindrical shell, spherical shell, solid cylinder, and cylindrical shell
2) The solid sphere, solid cylinder, cylindrical shell, and spherical shell.
3) The solid sphere, solid cylinder, spherical shell, and cylindrical shell.
4) The solid cylinder, solid sphere, spherical shell, and cylindrical shell.

Exercise

A ring and a circular disk, each with the mass and radius of M and R, are simultaneously freed from the top of an inclined surface. What is their arriving order?

Final Answer

Their arriving order is circular disk and ring.

9.7   Work-Energy Theorem

Problem

9.12. An electric motor with the power of $0.75\ hp$ rotates a wheel with the moment of inertia of $2\ kg.\ m^2$ for 8 seconds that was in the stationary state in the beginning. Calculate the angular velocity of the wheel if there is no energy waste (Fig. 9.13).

Difficulty level   ○ Easy   ● Normal   ○ Hard
Calculation amount   ○ Small   ● Normal   ○ Large

1) 34 rad/s
2) 49 rad/s
3) 55 rad/s
4) 67 rad/s

Problem

9.13. A 150 kg ring is rotating on horizontal surface. How much work is needed to stop it if the speed of its center of mass is 0.15 m/s?

Difficulty level   ○ Easy   ● Normal   ○ Hard
Calculation amount   ○ Small   ● Normal   ○ Large

1) 3.415 J
2) 3.125 J
3) 3.360 J
4) 3.375 J

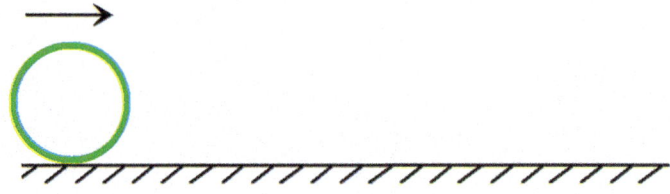

Fig. 9.13 The figure concerned with Problem 9.13

Partially Solved Exercise

In the previous problem, replace the ring with a circular disk with the same mass and velocity. Calculate the amount of work needed to stop it.

Solution

Based on the work-energy theorem, the net work done by the forces on an object equals the change in its kinetic energy. In other words:

$$W = \Delta K$$

In this problem, both linear and rotational kinetic energy exist. Therefore:

$$W = \left(\frac{1}{2}mv_{cm}^2 - \frac{1}{2}mv_{cm,0}^2\right) + \left(\frac{1}{2}I\omega^2 - \frac{1}{2}I\omega_0^2\right)$$

Additionally, the moment of inertia of a circular disk with the mass and radius of m and R around the axis, passing from its center and perpendicular to its base, can be calculated as follows.

$$I = (\qquad)$$

Hence:

$$W = ((\qquad) - (\qquad)) + ((\qquad) - (\qquad)) = (\qquad)$$

$$\Rightarrow W =$$

$$\Rightarrow |W| = 2.53\,J$$

References

1. Rahmani-Andebili, M. (2023). Calculus III – Practice Problems, Methods, and Solutions, Springer Nature.
2. Rahmani-Andebili, M. (2023). Calculus II – Practice Problems, Methods, and Solutions, Springer Nature.
3. Rahmani-Andebili, M. (2023). Calculus I (2nd Ed.) – Practice Problems, Methods, and Solutions, Springer Nature.
4. Rahmani-Andebili, M. (2024). Precalculus (2nd Ed.) – Practice Problems, Methods, and Solutions, Springer Nature.

Abstract

In this chapter, the problems of the ninth chapter are fully solved, in detail, step-by-step, and with different methods.

10.1 Rotational Kinematics

10.1. Since both wheels are connected to the same belt, their linear velocities are equal. However, based on the relation between the angular and linear velocities, their angular velocities are different [1–4].

$$v_1 = v_2$$

$$\Rightarrow r_1 \omega_1 = r_2 \omega_2$$

$$\Rightarrow \frac{\omega_2}{\omega_1} = \frac{r_1}{r_2} \Rightarrow \omega_2 < \omega_1$$

Choice (2) is the answer (Fig. 10.1).

Fig. 10.1 The figure concerned with Problem 10.1

10.2. Based on the information given in the problem, we have:

$$m = 2 \ kg$$

$$\vec{r}(t) = 3t^2 \widehat{i} + 2t \widehat{j}$$

As we know, the torque can be calculated as follows.

$$\vec{\tau} = \frac{d}{dt}\left(\vec{r} \times \vec{p}\right)$$

where,

$$\vec{p} = m\vec{v}$$

$$\vec{v} = \frac{d}{dt}\vec{r}$$

Hence:

$$\vec{\tau} = \frac{d}{dt}\left(\vec{r} \times m\frac{d}{dt}\vec{r}\right)$$

$$\Rightarrow \vec{\tau} = \frac{d}{dt}\left(\left(3t^2\widehat{i} + 2t\widehat{j}\right) \times 2\left(6t\widehat{i} + 2\widehat{j}\right)\right)$$

$$\Rightarrow \vec{\tau} = \frac{d}{dt}\left(12t^2\widehat{k} - 24t^2\widehat{k}\right) = \frac{d}{dt}\left(-12t^2\right)\widehat{k}$$

$$\Rightarrow \vec{\tau} = -24t\widehat{k}$$

Choice (3) is the answer.

> **Notes**
>
> In this problem, the relations below have been used.
>
> $$\frac{d}{dt}t^n = nt^{n-1}$$
>
> $$\left(a_1\widehat{i} + a_2\widehat{j}\right) \times \left(b_1\widehat{i} + b_2\widehat{j}\right) = (a_1b_2 - a_2b_1)\widehat{k}$$

10.2   Moment of Inertia

10.3. The moment of inertia of a system with discontinuous distribution of mass can be calculated as follows.

$$I = \sum_{i=1}^{N} m_i r_i^2$$

Herein, r_i is the distance of the i'th particle from the axis of rotation.

Therefore, for this system, we have:

$$I = m_1 r_1^2 + m_2 r_2^2 + m_3 r_3^2$$

$$\Rightarrow I = 5 \times 1^2 + 5 \times 1^2 + 5 \times 0^2$$

$$\Rightarrow I = 10 \ kg.m^2$$

Choice (2) is the answer (Fig. 10.2).

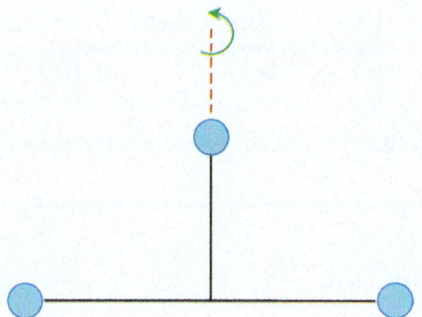

Fig. 10.2 The figure concerned with Problem 10.3

10.4. The moment of inertia of a body with continuous distribution of mass can be calculated as follows.

$$I = \int r^2 dm$$

Herein, r is the distance of the integration element dm from the axis of rotation.

Therefore, for this system, we have:

$$I = \int_0^L x^2 dm$$

$$dm = \lambda dx$$

$$\lambda = \frac{M}{L}$$

Thus:

$$I = \int_0^L x^2 \frac{M}{L} dx = \frac{M}{L} \int_0^L x^2 dx$$

$$\Rightarrow I = \frac{M}{L} \left[\frac{x^3}{3} \right]_0^L = \frac{M}{L} \frac{L^3}{3}$$

$$\Rightarrow I = \frac{1}{3} ML^2$$

Choice (2) is the answer (Fig. 10.3).

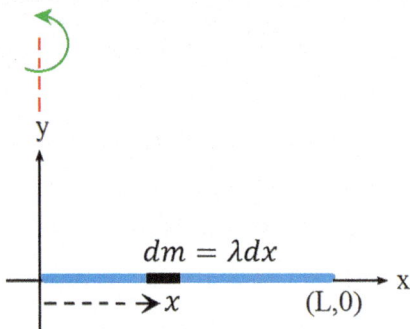

Fig. 10.3 The figure concerned with Problem 10.4

Notes

In this problem, the relation below has been used.

$$\int x^n dx = \frac{1}{n+1} x^{n+1} + c$$

10.3   Parallel Axis Theorem

10.5. Based on the information given in the problem, we have:

$$M = 5 \ kg$$

$$R = 0.1 \ m$$

$$m = 3 \ kg$$

$$L = 1 \ m$$

As we know, the moments of inertia of a solid sphere (with the mass and radius of M and R) around its diameter and a rod (with the mass and length of m and L) around the axis perpendicularly passing from one of its ends are as follows.

$$I_{\text{solid sphere}} = \frac{2}{5} MR^2$$

$$I_{\text{rod}} = \frac{1}{3} mL^2$$

As can be seen, the axis Δ is perpendicularly passing from one of the ends of the rod. However, the distance of the solid sphere from the axis is about the length of the rod (L).

Parallel axis theorem states if there is a new axis, which is parallel to the first axis but displaced from it by a distance L, the moment of inertia of a body with the mass of M around the new axis is calculated as follows.

$$I' = I + ML^2$$

Hence, we have:

$$I'_{\text{solid sphere}} = \frac{2}{5}MR^2 + M(L+R)^2$$

Therefore, the total moment of inertia of the system around the axis Δ can be calculated as follows.

$$I_{\text{total}} = I_{\text{rod}} + I'_{\text{solid sphere}}$$

$$\Rightarrow I_{\text{total}} = \frac{1}{3}mL^2 + \frac{2}{5}MR^2 + M(L+R)^2$$

$$\Rightarrow I_{\text{total}} = \frac{1}{3} \times 3 \times 1^2 + \frac{2}{5} \times 5 \times 0.1^2 + 5 \times (1+0.1)^2$$

$$\Rightarrow I_{\text{total}} = 1 + 0.02 + 6.05$$

$$\Rightarrow I = 7.07 \ kg.m^2$$

Choice (4) is the answer.

Note that if the radius of the solid sphere was ignored in the relation of the distance between the solid sphere and the axis, the misguiding answer of $6.02 \ kg. \ m^2$ would be achieved.

10.4 Rotational Dynamics

10.6. Based on the information given in the problem, we have:

$$M = 2 \ kg$$

$$m_1 = 3 \ kg$$

$$m_2 = 1 \ kg$$

$$g = 10 \ m/s^2$$

Since $m_1 > m_2$, the first mass starts moving down and the other one moves up.

Moreover, since they have been connected to a single string, their accelerations are the same. In other words:

$$a_1 = a_2 = a$$

In addition, since the pulley is not assumed weightless, the tension forces on its sides are not equal. In other words:

$$T_1 \neq T_2$$

Applying Newton's second law for m_1:

$$\sum F_1 = m_1 a$$

$$m_1 g - T_1 = m_1 a \qquad (10.1)$$

Applying Newton's second law for m_2:

$$\sum F_2 = m_2 a$$

$$T_2 - m_2 g = m_2 a \qquad (10.2)$$

Applying Newton's second law of rotation for the pulley:

$$(T_1 - T_2)r = I\alpha$$

Herein, r, I, and α are the radius, moment of inertia, and angular acceleration of the pulley, respectively. Since the pulley is a solid cylinder, its moment of inertial around its axis can be calculated as follows.

$$I = \frac{1}{2}Mr^2$$

In addition, as we know, the relation between the angular and linear acceleration is as follows.

$$a = r\alpha$$

Hence:

$$(T_1 - T_2)r = \frac{1}{2}Mr^2\frac{a}{r}$$

$$\Rightarrow T_1 - T_2 = \frac{1}{2}Ma \qquad (10.3)$$

Adding (10.1) and (10.2):

$$m_1 g - T_1 + T_2 - m_2 g = (m_1 + m_2)a$$

$$\Rightarrow g(m_1 - m_2) - (T_1 - T_2) = (m_1 + m_2)a \qquad (10.4)$$

Solving (10.3) and (10.4):

$$g(m_1 - m_2) - \frac{1}{2}Ma = (m_1 + m_2)a$$

$$\Rightarrow 10(3 - 1) - \frac{1}{2} \times 2a = (3 + 1)a$$

$$\Rightarrow 20 = 5a$$

$$\Rightarrow a = 4 \ m/s^2$$

Choice (3) is the answer (Fig. 10.4).

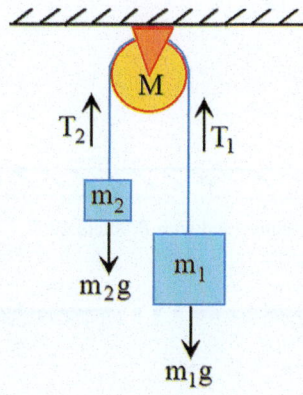

Fig. 10.4 The figure concerned with Problem 10.6

10.5 Rotational Kinetic Energy

10.7. The moment of inertia of a ring with the mass and radius of m and r around the axis, passing from its center and perpendicular to its base, can be calculated as follows.

$$I = mr^2$$

In addition, the relation between the angular and linear velocities is as follows.

$$v = r\omega$$

Therefore:

$$\frac{K_{rot}}{K_{tot}} = \frac{K_{rot}}{K_{lin} + K_{rot}} = \frac{\frac{1}{2}I\omega^2}{\frac{1}{2}mv_{cm}^2 + \frac{1}{2}I\omega^2}$$

$$\Rightarrow \frac{K_{rot}}{K_{tot}} = \frac{\frac{1}{2} \times mr^2\left(\frac{v_{cm}}{r}\right)^2}{\frac{1}{2}mv_{cm}^2 + \frac{1}{2} \times mr^2\left(\frac{v_{cm}}{r}\right)^2} = \frac{v_{cm}^2}{v_{cm}^2 + v_{cm}^2}$$

$$\Rightarrow \frac{K_{rot}}{K_{tot}} = \frac{1}{2}$$

Choice (1) is the answer (Fig. 10.5).

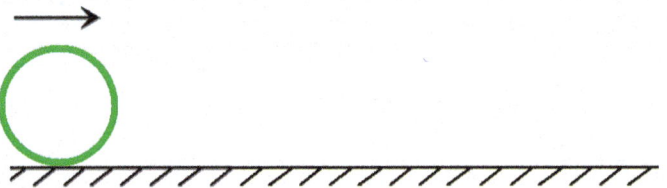

Fig. 10.5 The figure concerned with Problem 10.7

■ ■ ■

10.8. The moment of inertia of a sphere with mass and radius of m and r around the axis, passing from its diameter, can be calculated as follows.

$$I = \frac{2}{5}mr^2$$

In addition, the relation between the angular and linear velocities is as follows.

$$v = r\omega$$

Therefore:

$$\frac{K_{rot}}{K_{tot}} = \frac{K_{rot}}{K_{lin} + K_{rot}} = \frac{\frac{1}{2}I\omega^2}{\frac{1}{2}mv_{cm}^2 + \frac{1}{2}I\omega^2}$$

$$\Rightarrow \frac{K_{rot}}{K_{tot}} = \frac{\frac{1}{2} \times \frac{2}{5}mr^2\left(\frac{v_{cm}}{r}\right)^2}{\frac{1}{2}mv_{cm}^2 + \frac{1}{2} \times \frac{2}{5}mr^2\left(\frac{v_{cm}}{r}\right)^2} = \frac{\frac{2}{5}v_{cm}^2}{v_{cm}^2 + \frac{2}{5}v_{cm}^2} = \frac{\frac{2}{5}}{\frac{7}{5}}$$

$$\Rightarrow \frac{K_{rot}}{K_{tot}} = \frac{2}{7}$$

Choice (1) is the answer (Fig. 10.6).

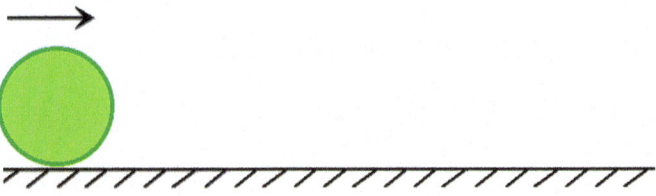

Fig. 10.6 The figure concerned with Problem 10.8

■ ■ ■

10.6 Conservation of Mechanical Energy Principle

10.9. Based on the information given in the problem, we have:

$$m = 2\ kg$$

$$R = 0.2\ m$$

The problem can be solved by applying the conservation of mechanical energy principle as follows.

$$E_1 = E_2$$

$$\Rightarrow K_1 + U_1 = K_2 + U_2$$

At the bottom of the inclined surface, the cylinder includes both linear and rotational kinetic energy. Thus:

$$mgh = \frac{1}{2}mv^2 + \frac{1}{2}I\omega^2$$

Moreover, the moment of inertia of a cylinder with the mass and radius of m and r around its axis can be calculated as follows.

$$I = \frac{1}{2}mr^2$$

In addition, the relation between the angular and linear velocities is as follows.

$$v = r\omega$$

Therefore:

$$m \times 10 \times 4 \sin 30° = \frac{1}{2}mv_{cm}^2 + \frac{1}{2} \times \frac{1}{2}mr^2 \times \left(\frac{v_{cm}}{r}\right)^2$$

$$\Rightarrow 20 = \frac{1}{2}v_{cm}^2 + \frac{1}{4}v_{cm}^2$$

$$\Rightarrow 20 = \frac{3}{4}v_{cm}^2 \Rightarrow v_{cm}^2 = \frac{80}{3}$$

$$\Rightarrow v_{cm} = \sqrt{\frac{80}{3}}\ m/s$$

Choice (2) is the answer (Fig. 10.7).

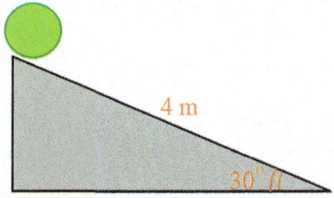

Fig. 10.7 The figure concerned with Problem 10.9

Notes
In this problem, the relation below has been used.

$$\sin 30° = \frac{1}{2}$$

10.10. The problem can be solved by applying the conservation of mechanical energy principle as follows.

$$E_1 = E_2$$

$$\Rightarrow Mgh = \frac{1}{2}I\omega^2 + \frac{1}{2}Mv_{cm}^2$$

As we know, the relation between the angular and linear velocities is as follows.

$$v = R\omega$$

In addition, the moment of inertia of a ring with the mass and radius of M and R around the axis, passing from its center and perpendicular to its base, is as follows.

$$I_{ring} = MR^2$$

Hence:

$$Mgh = \frac{1}{2}MR^2\omega^2 + \frac{1}{2}M(R\omega)^2$$

$$\Rightarrow gh = R^2\omega^2$$

$$\Rightarrow \omega = \frac{1}{R}\sqrt{gh}$$

On the other hand, we know that the angular momentum can be calculated as follows:

$$L = I\omega$$

Thus:

$$L = MR^2 \times \frac{1}{R}\sqrt{gh}$$

$$\Rightarrow L = MR\sqrt{gh}$$

Choice (1) is the answer (Fig. 10.8).

Fig. 10.8 The figure concerned with Problem 10.10

10.11. The problem can be solved by applying the conservation of mechanical energy principle as follows.

$$E_1 = E_2$$

$$\Rightarrow Mgh = \frac{1}{2}I\omega^2 + \frac{1}{2}Mv^2$$

As we know, the relation between the angular and linear velocities is as follows.

$$\omega = \frac{v}{R}$$

In addition, the moments of inertia of a cylindrical shell around its axis, a solid cylinder around its axis, a spherical shell around its diameter, and a solid sphere around its diameter, each with the mass and radius of M and R, are as follows.

$$I_{\text{cylindrical shell}} = MR^2$$

$$I_{\text{solid cylinder}} = \frac{1}{2}MR^2$$

$$I_{\text{spherical shell}} = \frac{2}{3}MR^2$$

$$I_{\text{solid sphere}} = \frac{2}{5}MR^2$$

Therefore, the conservation of mechanical energy principle for the cylindrical shell is as follows.

$$Mgh = \frac{1}{2}MR^2 \left(\frac{v_{\text{cylindrical shell}}}{R}\right)^2 + \frac{1}{2}Mv^2_{\text{cylindrical shell}}$$

$$\Rightarrow gh = \frac{1}{2}v^2_{\text{cylindrical shell}} + \frac{1}{2}v^2_{\text{cylindrical shell}}$$

$$\Rightarrow v_{\text{cylindrical shell}} = \sqrt{gh}$$

The conservation of mechanical energy principle for the solid cylinder is as follows.

$$Mgh = \frac{1}{2} \times \frac{1}{2}MR^2 \left(\frac{v_{\text{solid cylinder}}}{R}\right)^2 + \frac{1}{2}Mv^2_{\text{solid cylinder}}$$

$$\Rightarrow gh = \frac{1}{4}v^2_{\text{solid cylinder}} + \frac{1}{2}v^2_{\text{solid cylinder}}$$

$$\Rightarrow v_{\text{solid cylinder}} = \frac{2}{\sqrt{3}}\sqrt{gh}$$

The conservation of mechanical energy principle for the spherical shell is as follows.

$$Mgh = \frac{1}{2} \times \frac{2}{3}MR^2\left(\frac{v_{\text{spherical shell}}}{R}\right)^2 + \frac{1}{2}Mv^2_{\text{spherical shell}}$$

$$\Rightarrow gh = \frac{1}{3}v^2_{\text{spherical shell}} + \frac{1}{2}v^2_{\text{spherical shell}}$$

$$\Rightarrow v_{\text{spherical shell}} = \sqrt{\frac{6}{5}}\sqrt{gh}$$

The conservation of mechanical energy principle for the solid sphere is as follows.

$$Mgh = \frac{1}{2} \times \frac{2}{5}MR^2\left(\frac{v_{\text{solid sphere}}}{R}\right)^2 + \frac{1}{2}Mv^2_{\text{solid sphere}}$$

$$\Rightarrow gh = \frac{1}{5}v^2_{\text{solid sphere}} + \frac{1}{2}v^2_{\text{solid sphere}}$$

$$\Rightarrow v_{\text{solid sphere}} = \sqrt{\frac{10}{7}}\sqrt{gh}$$

Since $\sqrt{\frac{10}{7}} > \frac{2}{\sqrt{3}} > \sqrt{\frac{6}{5}} > 1$, we can conclude that:

$$v_{\text{solid sphere}} > v_{\text{solid cylinder}} > v_{\text{spherical shell}} > v_{\text{cylindrical shell}}$$

Hence, their arriving order is solid sphere, solid cylinder, spherical shell, and cylindrical shell. Choice (3) is the answer.

■ ■ ■

10.7   Work-Energy Theorem

10.12. Based on the information given in the problem, we have:

$$P = 0.75 \ hp$$

$$I = 2 \ kg.m^2$$

$$\Delta t = 8 \ s$$

$$\omega_0 = 0 \ rad/s$$

The work-energy theorem states that the net work done by the forces on an object equals the change in its kinetic energy. In other words:

$$W = \Delta K$$

In this problem, only rotational kinetic energy exists. Thus:

$$W = \frac{1}{2} I \omega^2 - \frac{1}{2} I \omega_0{}^2$$

On the other hand, we know that:

$$W = P \Delta t$$

Therefore:

$$P \Delta t = \frac{1}{2} I \omega^2 - \frac{1}{2} I \omega_0{}^2$$

$$\Rightarrow 0.75 \times 746 \times 8 = \frac{1}{2} \times 2 \times \omega^2$$

$$\Rightarrow \omega = \sqrt{4476} = 67 \; rad/s$$

Choice (4) is the answer.

Notes

In this problem, the relation below has been used.

$$1 \; hp = 746 \; J$$

■ ■ ■

10.13. Based on the information given in the problem, we have:

$$m = 150 \; kg$$

$$v_{cm,0} = 0.15 \; m/s$$

$$v_{cm} = 0 \; m/s$$

$$\omega = 0 \; rad/s$$

From the work-energy theorem, we know that the net work done by the forces on an object equals the change in its kinetic energy. In other words:

$$W = \Delta K$$

In this problem, both linear and rotational kinetic energy exist. Hence:

$$W = \left(\frac{1}{2}mv_{cm}^2 - \frac{1}{2}mv_{cm,0}^2\right) + \left(\frac{1}{2}I\omega^2 - \frac{1}{2}I\omega_0{}^2\right)$$

In addition, as we know, the moment of inertia of a ring with the mass and radius of m and R around the axis, passing from its center and perpendicular to its base, can be calculated as follows.

$$I = mR^2$$

Also, the relation between the angular and linear velocities is as follows.

$$v = R\omega$$

Therefore:

$$W = \left(0 - \frac{1}{2}mv_{cm,0}^2\right) + \left(0 - \frac{1}{2}mR^2\omega_0{}^2\right) = -\frac{1}{2}mv_{cm,0}^2 - \frac{1}{2}mv_{cm,0}^2 = -mv_{cm,0}^2$$

$$\Rightarrow W = -150 \times 0.15^2 = -3.375 \, J$$

$$\Rightarrow |W| = 3.375 \, J$$

Choice (4) is the answer (Fig. 10.9).

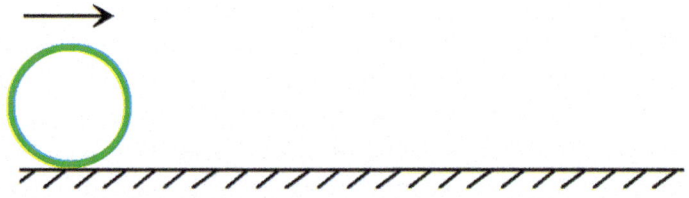

Fig. 10.9   A ring rotating on a horizontal surface with constant velocity

References

1. Rahmani-Andebili, M. (2023). Calculus III – Practice Problems, Methods, and Solutions, Springer Nature.
2. Rahmani-Andebili, M. (2023). Calculus II – Practice Problems, Methods, and Solutions, Springer Nature.
3. Rahmani-Andebili, M. (2023). Calculus I (2nd Ed.) – Practice Problems, Methods, and Solutions, Springer Nature.
4. Rahmani-Andebili, M. (2024). Precalculus (2nd Ed.) – Practice Problems, Methods, and Solutions, Springer Nature.

Simple Harmonic Motion: Part A

11

Abstract

In this chapter, the basic and advanced problems of simple harmonic motion are studied. The subjects include equations of motion, velocity, acceleration, and force of a simple harmonic oscillator; kinetic, potential, and mechanical energy of a simple harmonic motion; and period and frequency of a simple harmonic oscillator. Herein, different types of problems and exercises are presented that are categorized as follows.

- *Problems with detailed solution*: They have been designed to teach students the subjects in detail. Moreover, they have been categorized in different levels based on their difficulty levels (easy, normal, and hard) and calculation amounts (small, normal, and large).
- *Partially solved exercises*: They have been designed to encourage students to practice problems while guiding them through the problem-solving procedure and hinting the required formulas.
- *Exercises with final answer*: They have been designed to encourage students to practice more by themselves while hinting them by the final answer as well as to help instructors to give tests or quizzes.

11.1 Equations of Motion, Velocity, Acceleration, and Force of a Simple Harmonic Oscillator

Problem

11.1. The position-time curve of a simple harmonic oscillator is shown in Fig. 11.1. Calculate its equation of motion [1–4].

Difficulty level ○ Easy ● Normal ○ Hard
Calculation amount ○ Small ● Normal ○ Large

1) $y(t) = 4 \sin\left(12.5\pi t + \dfrac{\pi}{6}\right)$

2) $y(t) = 4 \sin\left(12.5\pi t + \dfrac{\pi}{3}\right)$

3) $y(t) = 2 \sin(12.5\pi t)$

4) $y(t) = 4 \sin(12.5\pi t)$

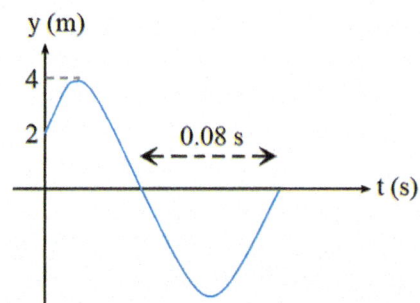

Fig. 11.1 The figure concerned with Problem 11.1

© The Author(s), under exclusive license to Springer Nature Switzerland AG 2025
M. Rahmani-Andebili, *General Physics I*, https://doi.org/10.1007/978-3-031-92862-8_11

Exercise

Calculate the equation of motion of a simple harmonic oscillator with the position-time curve illustrated in Fig. 11.2.

Final Answer

$$y(t) = 2\sin\left(t + \frac{\pi}{3}\right)$$

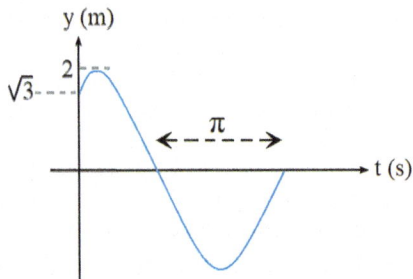

Fig. 11.2 The figure concerned with the exercise

Problem

11.2. The equation of motion of a simple harmonic oscillator is $x(t) = 2\sin\left(2\pi t + \frac{\pi}{6}\right)$. How long after $t = 0\ s$, the velocity of the object will be maximum for the first time.

Difficulty level   ○ Easy   ● Normal   ○ Hard
Calculation amount   ○ Small   ● Normal   ○ Large

1) $\frac{5}{7} s$

2) $\frac{7}{12} s$

3) $\frac{1}{12} s$

4) $\frac{5}{12} s$

Partially Solved Exercise

In the previous problem, determine the time (after $t = 0\ s$) that the acceleration of the object will be maximum for the second time.

Solution

The equation of acceleration of the simple harmonic oscillator can be calculated as follows.

$$a(t) = \frac{d}{dt}v(t) = \frac{d^2}{dt^2}x(t)$$

$$\Rightarrow a(t) = (\quad)\sin(\qquad)$$

The function is maximum when the sine is at its maximum magnitude, that is, ± 1. Therefore:

$$2\pi t + \frac{\pi}{6} = (\quad),(\quad)$$

$$\Rightarrow 2\pi t = (\quad)\sin(\quad)$$

$$\Rightarrow t = \frac{1}{6}, \frac{2}{6} s$$

As can be noticed, $t = \frac{2}{3} s$ is the time that the acceleration of the object will be maximum for the second time.

Problem

11.3. The maximum acceleration and maximum velocity of a simple harmonic oscillator are $100 \frac{m}{s^2}$ and $4 \frac{m}{s}$, respectively. Calculate the period of the oscillator in seconds.

Difficulty level ○ Easy ● Normal ○ Hard
Calculation amount ○ Small ● Normal ○ Large
1) 0.08π
2) 0.04π
3) 0.02π
4) 0.01π

Exercise

Calculate the frequency of a simple harmonic oscillator that its maximum acceleration and maximum domain are $100 \frac{m}{s^2}$ and $1\ m$, respectively.

Final Answer

$f = \frac{50}{\pi} Hz$

Problem

11.4. The equation of motion of a simple harmonic oscillator is $x(t) = 0.1 \sin\left(\pi t + \frac{\pi}{6}\right)$. Calculate the force exerted on the 1 kg object at $t = 1\ s$. Herein, assume that $\pi^2 = 10$.

Difficulty level ○ Easy ● Normal ○ Hard
Calculation amount ○ Small ● Normal ○ Large
1) $0.25\ N$
2) $0.5\ N$
3) $0.125\ N$
4) $0.1\ N$

Exercise

Calculate the magnitude of the force exerted on a 1 kg simple harmonic oscillator at $t = 0.5\ s$ that its equation of velocity is $(t) = \cos(\pi t)$.

Final Answer

$F = \pi\ N$

Problem

11.5. What is the magnitude of the ratio of the velocity of an oscillator to its maximum velocity when its position is at 50% of its amplitude?

Difficulty level ○ Easy ● Normal ○ Hard
Calculation amount ○ Small ● Normal ○ Large

1) $\frac{\sqrt{3}}{6}$
2) $\frac{\sqrt{3}}{2}$

3) $\dfrac{1}{4}$

4) $\dfrac{1}{2}$

Problem

11.6. Calculate the magnitude of the ratio of the acceleration of an oscillator to its maximum acceleration when its velocity is at 50% of its maximum velocity.

Difficulty level   ○ Easy   ● Normal   ○ Hard

Calculation amount   ○ Small   ● Normal   ○ Large

1) $\dfrac{\sqrt{2}}{2}$

2) $\dfrac{\sqrt{3}}{2}$

3) 1

4) $\dfrac{1}{2}$

Exercise

Calculate the magnitude of the ratio of the velocity of an oscillator to its maximum velocity when its position is $\dfrac{\sqrt{3}}{2}$ times of its amplitude.

Final Answer

$$\left| \dfrac{v}{v_{max}} \right| = 0.5$$

Problem

11.7. Figure 11.3 shows a simple harmonic oscillator which is in its equilibrium state. If the 0.2 kg mass is pulled down about 5 cm, and then freed, calculate the magnitude of velocity of the mass when it is 1 cm above its equilibrium position. The stiffness constant of the spring is 180 N/m.

Difficulty level   ○ Easy   ● Normal   ○ Hard

Calculation amount   ○ Small   ● Normal   ○ Large

1) 0.9 *m/s*

2) 0.5 *m/s*

3) 0.8 *m/s*

4) 0.4 *m/s*

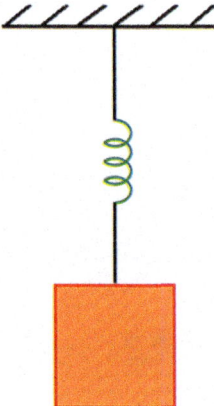

Fig. 11.3 The figure concerned with Problem 11.7

Partially Solved Exercise

In the previous problem, calculate the magnitude of acceleration of the mass.

Solution

As we know, the relation between the instantaneous acceleration and velocity of a simple harmonic oscillator is as follows

$$a = \pm \omega \sqrt{v_{max}^2 - v^2} = \pm \omega \sqrt{(A\omega)^2 - v^2}$$

From Problem 11.7, we have:

$$A = 0.05 \ m$$

$$\omega = 30 \ rad/s$$

$$|v| = 0.9 \ m/s$$

Therefore:

$$a = \pm (\quad) \sqrt{(\quad)^2 - (\quad)^2} = \pm (\quad)$$

$$\Rightarrow |a| = 36 \ m/s^2$$

11.2 Kinetic, Potential, and Mechanical Energy of a Simple Harmonic Oscillator

Problem

11.8. A body has a simple harmonic oscillation. Calculate the ratio of its kinetic energy to its total mechanical energy when its position is at 50% of its amplitude.

Difficulty level   ○ Easy   ● Normal   ○ Hard
Calculation amount   ○ Small ● Normal   ○ Large

1) $\frac{1}{4}$
2) $\frac{1}{2}$
3) $\frac{3}{4}$
4) 1

Partially Solved Exercise

In a simple harmonic motion, the position of the object is at 50% of its amplitude. Calculate the ratio of the kinetic energy of the object to its potential energy.

Solution

Based on the information given in the problem, we have:

$$x(t) = \frac{A}{2}$$

The ratio of kinetic energy to potential energy can be calculated as follows.

$$\frac{K}{U} = \frac{E - U}{U} = \frac{E}{U} - 1$$

$$\Rightarrow \frac{K}{U} = \frac{(\qquad)}{(\qquad)} - 1$$

For $x = \frac{A}{2}$, we have:

$$\Rightarrow \frac{K}{U} = \frac{(\qquad)}{(\qquad)} - 1$$

$$\Rightarrow \frac{K}{U} = 3$$

Notes

In this problem, the relations below have been used.

$$E = U + K$$

$$K = \frac{1}{2}kx^2$$

$$E = \frac{1}{2}kA^2$$

Problem

11.9. The maximum velocity of a 0.1 kg object in the simple harmonic motion is 10π m/s. Calculate its total mechanical energy. Herein, assume that $\pi^2 = 10$.

Difficulty level   ○ Easy   ○ Normal   ● Hard
Calculation amount   ○ Small   ● Normal   ○ Large

1) 50 J
2) 100 J
3) 200 J
4) 10 J

11.3   Period and Frequency of a Simple Harmonic Oscillator

Problem

11.10. An object with the mass of m that has been connected to a spring oscillates with the period of T_1. If the spring is halved, one part of the spring is used, and the object is replaced with another object with the mass of $2m$, the period of the oscillator becomes T_2. Calculate the value of $\frac{T_2}{T_1}$

Difficulty level   ○ Easy   ○ Normal   ● Hard
Calculation amount   ○ Small   ● Normal   ○ Large

1) 4
2) 0.5
3) 2
4) 1

Exercise

Solve the previous problem by assuming that, after halving the spring, the two parts are connected in parallel.

Final Answer

$$\frac{T_2}{T_1} = \frac{\sqrt{2}}{2}$$

Problem

11.11. When a pendulum oscillates 15 times, the other one oscillates 10 times. Calculate the ratio of lengths of these two pendulums.

Difficulty level   ○ Easy   ○ Normal   ● Hard
Calculation amount   ○ Small   ● Normal   ○ Large

1) 0.11
2) 0.22
3) 0.33
4) 0.44

Exercise

Calculate the ratio of frequency of two pendulums in which $l_2 = 4l_1$.

Final Answer

$$\frac{f_2}{f_1} = 0.5$$

Problem

11.12. In the system shown in Fig. 11.4, calculate the frequency of the oscillations.

Difficulty level   ○ Easy   ○ Normal   ● Hard
Calculation amount   ○ Small   ● Normal   ○ Large

1) $\dfrac{1}{2\pi} \sqrt{\dfrac{k_1 + k_2}{(k_1 + k_2)M}}$

2) $\dfrac{1}{2\pi} \sqrt{\dfrac{M}{k_1 + k_2}}$

3) $\dfrac{1}{2\pi} \sqrt{\dfrac{k_1 + k_2}{M}}$

4) $\dfrac{1}{2\pi} \sqrt{\dfrac{(k_1 + k_2)M}{k_1 + k_2}}$

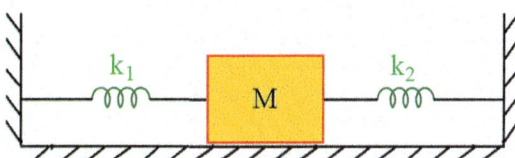

Fig. 11.4 The figure concerned with Problem 11.12

Problem

11.13. In the system shown in Fig. 11.5, calculate the period of the oscillations.

Difficulty level   ○ Easy   ○ Normal   ● Hard
Calculation amount   ○ Small   ● Normal   ○ Large

1) $2\pi\sqrt{\dfrac{M(k_1 + k_2)}{k_1 k_2}}$

2) $2\pi\sqrt{\dfrac{Mk_1 k_2}{k_1 + k_2}}$

3) $\dfrac{1}{2\pi}\sqrt{\dfrac{M(k_1 + k_2)}{k_1 k_2}}$

4) $\dfrac{1}{2\pi}\sqrt{\dfrac{Mk_1 k_2}{k_1 + k_2}}$

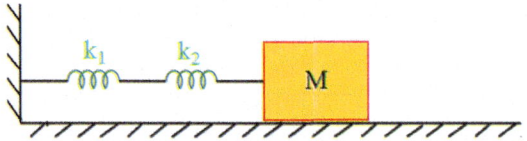

Fig. 11.5 The figure concerned with Problem 11.13

Exercise

In the system shown in Fig. 11.6, calculate the period of the oscillations.

Final Answer

$$T = 2\pi\sqrt{\frac{m}{k_1 + k_2}}$$

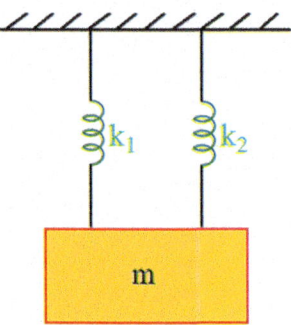

Fig. 11.6 The figure concerned with the exercise

References

1. Rahmani-Andebili, M. (2023). Calculus III – Practice Problems, Methods, and Solutions, Springer Nature.
2. Rahmani-Andebili, M. (2023). Calculus II – Practice Problems, Methods, and Solutions, Springer Nature.
3. Rahmani-Andebili, M. (2023). Calculus I (2nd Ed.) – Practice Problems, Methods, and Solutions, Springer Nature.
4. Rahmani-Andebili, M. (2024). Precalculus (2nd Ed.) – Practice Problems, Methods, and Solutions, Springer Nature.

Abstract

In this chapter, the problems of the eleventh chapter are fully solved, in detail, step-by-step, and with different methods.

12.1 Equations of Motion, Velocity, Acceleration, and Force of a Simple Harmonic Oscillator

12.1. The general form of the position-time function of a simple harmonic oscillator is as follows [1–4].

$$y(t) = A \sin(\omega t + \theta_0)$$

where, A, ω, and θ_0 are the amplitude, angular frequency, and initial phase angle of the oscillator, respectively.

The information below is noticed from Fig. 12.1.

$$A = 4 \ m$$

$$\frac{T}{2} = 0.08 \Rightarrow T = 0.16 \ s \Rightarrow \omega = \frac{2\pi}{T} = \frac{2\pi}{0.16} = 12.5\pi \ rad/s$$

$$\sin \theta_0 = \frac{2}{4} \Rightarrow \theta_0 = \sin^{-1} \frac{1}{2} = \frac{\pi}{6} \ rad$$

Therefore:

$$y(t) = 4 \sin\left(12.5\pi t + \frac{\pi}{6}\right)$$

Choice (1) is the answer.

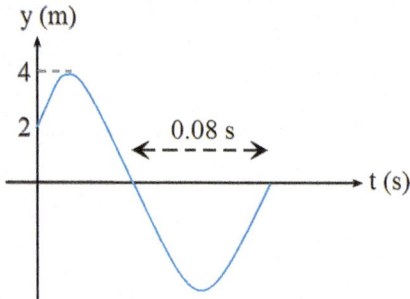

Fig. 12.1 The figure concerned with Problem 12.1

12.2. Based on the information given in the problem, we have:

$$x(t) = 2\sin\left(2\pi t + \frac{\pi}{6}\right)$$

The equation of velocity of the simple harmonic oscillator can be calculated as follows.

$$v(t) = \frac{d}{dt}x(t)$$

$$\Rightarrow v(t) = 4\pi\cos\left(2\pi t + \frac{\pi}{6}\right)$$

The function is maximum when the cosine is at its maximum magnitude, that is, $\cos\left(2\pi t + \frac{\pi}{6}\right) = \pm 1$. Therefore:

$$2\pi t + \frac{\pi}{6} = 0, \pi$$

$$\Rightarrow 2\pi t = -\frac{\pi}{6}, \frac{5\pi}{6}$$

$$\Rightarrow t = -\frac{1}{12}, \frac{5}{12}\ s$$

However, $t = -\frac{1}{12}$ s is not acceptable because it is not after $t = 0$. Hence:

$$\Rightarrow t = \frac{5}{12}\ s$$

Choice (4) is the answer.

Notes

In this problem, the relations below have been used.

$$\cos^{-1}1 = 0$$

$$\cos^{-1}(-1) = \pi$$

12.3. Based on the information given in the problem, we have:

$$a_{max} = 100 \; m/s^2$$

$$v_{max} = 4 \; m/s$$

The maximum acceleration and maximum velocity of a simple harmonic oscillator can be calculated as follows.

$$a_{max} = A\omega^2$$

$$v_{max} = A\omega$$

Thus:

$$\frac{a_{max}}{v_{max}} = \frac{A\omega^2}{A\omega} \Rightarrow \frac{a_{max}}{v_{max}} = \omega$$

$$\Rightarrow \frac{100}{4} = \omega \Rightarrow \omega = 25 \; rad/s$$

$$\Rightarrow \frac{2\pi}{T} = 25 \Rightarrow T = \frac{2\pi}{25}$$

$$\Rightarrow T = 0.08\pi \; s$$

Choice (1) is the answer.

12.4. Based on the information given in the problem, we have:

$$x(t) = 0.1 \sin\left(\pi t + \frac{\pi}{6}\right)$$

$$m = 1 \; kg$$

$$t = 1 \; s$$

$$\pi^2 = 10$$

The equation of velocity of the simple harmonic oscillator can be calculated as follows.

$$v(t) = \frac{d}{dt}x(t)$$

$$\Rightarrow v(t) = 0.1\pi \cos\left(\pi t + \frac{\pi}{6}\right)$$

Moreover, the equation of acceleration of the simple harmonic oscillator can be calculated as follows.

$$a(t) = \frac{d}{dt}v(t) = \frac{d^2}{dt^2}x(t)$$

$$\Rightarrow a(t) = -0.1\pi^2 \sin\left(\pi t + \frac{\pi}{6}\right)$$

Additionally, the equation of force of the simple harmonic oscillator can be calculated as follows.

$$F(t) = ma(t)$$

$$\Rightarrow F(t) = -0.1\pi^2 \sin\left(\pi t + \frac{\pi}{6}\right)$$

At $t = 1$ s, we have:

$$F(t=1) = -0.1\pi^2 \sin\left(\frac{7\pi}{6}\right)$$

$$\Rightarrow F(t=1) = -0.1 \times 10\left(-\sin\frac{\pi}{6}\right)$$

$$\Rightarrow F(t=1) = 0.5\ N$$

Choice (2) is the answer.

Notes

In this problem, the relations below have been used.

$$\sin(\pi + \theta) = -\sin\theta$$

$$\sin\frac{\pi}{6} = \frac{1}{2}$$

12.5. Based on the information given in the problem, we have:

$$x(t) = \frac{A}{2} \tag{12.1}$$

Method 1 As we know, the general form of the equation of position of a simple harmonic oscillator is as follows.

$$x(t) = A \sin(\omega t + \theta_0) = A \sin(\theta(t)) \tag{12.2}$$

The equation of velocity of the simple harmonic oscillator can be calculated as follows.

$$v(t) = \frac{d}{dt}x(t) = A\omega \cos(\theta(t)) \tag{12.3}$$

Solving (12.1) and (12.2):

$$A \sin(\theta(t)) = \frac{A}{2} \Rightarrow \sin(\theta(t)) = \frac{1}{2}$$

$$\theta(t) = \frac{\pi}{6} \qquad (12.4)$$

Solving (12.3) and (12.4):

$$v(t) = A\omega \frac{\sqrt{3}}{2}$$

$$\Rightarrow \left| \frac{v}{v_{max}} \right| = \frac{\left| A\omega \frac{\sqrt{3}}{2} \right|}{A\omega} \Rightarrow \left| \frac{v}{v_{max}} \right| = \frac{\sqrt{3}}{2}$$

Choice (2) is the answer.

Method 2 The problem can be solved by using the equation below that shows the relation between the instantaneous velocity and position of a simple harmonic oscillator.

$$v = \pm \omega \sqrt{A^2 - x^2}$$

For $x = \frac{A}{2}$, we have:

$$v = \pm \omega \sqrt{A^2 - \left(\frac{A}{2} \right)^2}$$

$$\Rightarrow v = \pm \frac{\sqrt{3}}{2} A\omega$$

Therefore:

$$\left| \frac{v}{v_{max}} \right| = \frac{\left| \pm \frac{\sqrt{3}}{2} A\omega \right|}{A\omega} \Rightarrow \left| \frac{v}{v_{max}} \right| = \frac{\sqrt{3}}{2}$$

Notes

In this problem, the relations below have been used.

$$\sin^{-1} \frac{1}{2} = \frac{\pi}{6}$$

$$\cos \frac{\pi}{3} = \frac{\sqrt{3}}{2}$$

$$v_{max} = A\omega$$

12.6. Based on the information given in the problem, we have:

$$v(t) = \frac{\sqrt{2}}{2} v_{max} \qquad (12.5)$$

The problem can be solved by using the equation below that shows the relation between the instantaneous acceleration and velocity of a simple harmonic oscillator.

$$a = \pm \omega \sqrt{v_{max}^2 - v^2}$$

For $v = \frac{\sqrt{2}}{2} v_{max}$, we have:

$$a = \pm \omega \sqrt{v_{max}^2 - \left(\frac{\sqrt{2}}{2} v_{max}\right)^2}$$

$$\Rightarrow a = \pm \frac{\sqrt{2}}{2} v_{max} \omega$$

Therefore:

$$\left|\frac{a}{a_{max}}\right| = \left|\frac{\pm \frac{\sqrt{2}}{2}(A\omega)\omega}{A\omega^2}\right| \Rightarrow \left|\frac{a}{a_{max}}\right| = \frac{\sqrt{2}}{2}$$

Choice (1) is the answer.

Notes

In this problem, the relations below have been used.

$$v_{max} = A\omega$$

$$a_{max} = A\omega^2$$

12.7. Based on the information given in the problem, we have:

$$m = 0.2 \ kg$$

$$A = 0.05 \ m$$

$$x = 0.04 \ m$$

$$k = 180 \ N/m$$

The relation between the angular velocity and stiffness constant of a spring is as follows.

$$k = m\omega^2$$

Hence:

$$180 = 0.2\omega^2 \Rightarrow \omega^2 = 900 \Rightarrow \omega = 30 \ rad/s$$

As we know, the relation between the instantaneous velocity and position of a simple harmonic oscillator is as follows

$$v = \pm \omega \sqrt{A^2 - x^2}$$

Therefore:

$$v = \pm 30\sqrt{0.05^2 - 0.04^2} = \pm 30\sqrt{0.0009}$$

$$\Rightarrow |v| = 0.9 \, m/s$$

Choice (1) is the answer (Fig. 12.2).

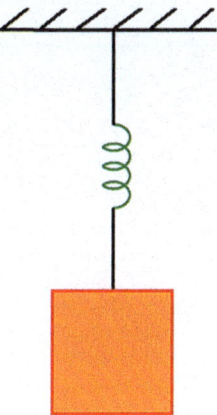

Fig. 12.2 The figure concerned with Problem 12.7

12.2 Kinetic, Potential, and Mechanical Energy of a Simple Harmonic Oscillator

12.8. Based on the information given in the problem, we have:

$$x(t) = \frac{A}{2}$$

The ratio of kinetic energy to total mechanical energy can be calculated as follows.

$$\frac{K}{E} = \frac{E - U}{E} = 1 - \frac{U}{E}$$

$$\Rightarrow \frac{K}{E} = 1 - \frac{\frac{1}{2}kx^2}{\frac{1}{2}kA^2}$$

For $x = \dfrac{A}{2}$, we have:

$$\Rightarrow \frac{K}{E} = 1 - \frac{\frac{1}{2}k\left(\frac{A}{2}\right)^2}{\frac{1}{2}kA^2}$$

$$\Rightarrow \frac{K}{E} = 1 - \frac{1}{4}$$

$$\Rightarrow \frac{K}{E} = \frac{3}{4}$$

Choice (3) is the answer.

> **Notes**
>
> In this problem, the relations below have been used.
>
> $$E = U + K$$
>
> $$K = \frac{1}{2}kx^2$$
>
> $$E = \frac{1}{2}kA^2$$

■ ■ ■

12.9. Based on the information given in the problem, we have:

$$m = 0.1 \; kg$$

$$v_{max} = 10\pi \; m/s$$

$$\pi^2 = 10$$

When a simple harmonic oscillator is at its maximum velocity, its potential energy is zero. In other words, the mechanical energy of the oscillator is equal to its maximum kinetic energy, that is:

$$E = K_{max}$$

$$\Rightarrow E = \frac{1}{2}mv_{max}^2$$

$$\Rightarrow E = \frac{1}{2} \times 0.1(10\pi)^2 = \frac{1}{2} \times 0.1 \times 100 \times 10$$

$$\Rightarrow E = 50 \; J$$

Choice (1) is the answer.

■ ■ ■

12.3   Period and Frequency of a Simple Harmonic Oscillator

12.10. Based on the information given in the problem, we have:

$$m_2 = 2m_1$$

If a spring is halved, the stiffness constant of each part will double. Hence:

$$k_2 = 2k_1$$

On the other hand, the period of the oscillations of a spring can be calculated as follows.

$$T = 2\pi\sqrt{\frac{m}{k}}$$

Hence, for this problem we can write:

$$\Rightarrow \frac{T_2}{T_1} = \sqrt{\frac{m_2}{m_1} \times \frac{k_1}{k_2}}$$

$$\Rightarrow \frac{T_2}{T_1} = \sqrt{\frac{2m}{m} \times \frac{k_1}{2k_1}}$$

$$\Rightarrow \frac{T_2}{T_1} = 1$$

Choice (4) is the answer.

12.11. The period of oscillations of a simple pendulum is as follows.

$$T = 2\pi\sqrt{\frac{l}{g}}$$

Therefore, the ratio of periods of the two pendulums can be calculated as follows.

$$\frac{T_2}{T_1} = \sqrt{\frac{l_2}{l_1}}$$

$$\Rightarrow \frac{\frac{1}{15}}{\frac{1}{10}} = \sqrt{\frac{l_2}{l_1}} \Rightarrow \frac{10}{15} = \sqrt{\frac{l_2}{l_1}} \Rightarrow \sqrt{\frac{l_2}{l_1}} = \frac{2}{3}$$

$$\Rightarrow \frac{l_2}{l_1} = 0.44$$

Choice (4) is the answer.

12.12. It should be noted that the two springs in Fig. 12.3 are connected in parallel. Thus, the total stiffness constant of the springs is as follows.

$$k_{tot} = k_1 + k_2$$

On the other hand, we know that the frequency of the oscillations of a simple spring can be calculated as follows.

$$f = \frac{1}{2\pi}\sqrt{\frac{k}{M}}$$

Therefore:

$$f = \frac{1}{2\pi}\sqrt{\frac{k_1 + k_2}{M}}$$

Choice (3) is the answer.

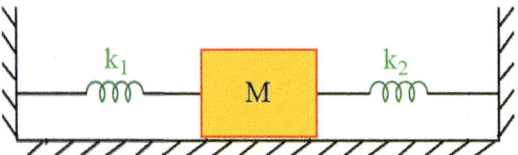

Fig. 12.3 The figure concerned with Problem 12.12

12.13. The two springs in Fig. 12.4 are connected in series. Hence, the total stiffness constant of the springs is as follows.

$$k_{tot} = \frac{k_1 k_2}{k_1 + k_2}$$

On the other hand, we know that the period of the oscillations of a simple spring can be calculated as follows

$$T = 2\pi\sqrt{\frac{M}{k}}$$

Therefore:

$$T = 2\pi\sqrt{\frac{M(k_1 + k_2)}{k_1 k_2}}$$

Choice (1) is the answer.

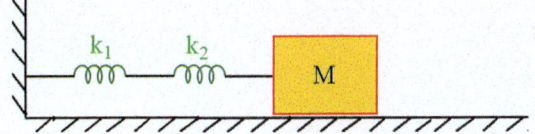

Fig. 12.4 The figure concerned with Problem 12.13

References

1. Rahmani-Andebili, M. (2023). Calculus III – Practice Problems, Methods, and Solutions, Springer Nature.
2. Rahmani-Andebili, M. (2023). Calculus II – Practice Problems, Methods, and Solutions, Springer Nature.
3. Rahmani-Andebili, M. (2023). Calculus I (2nd Ed.) – Practice Problems, Methods, and Solutions, Springer Nature.
4. Rahmani-Andebili, M. (2024). Precalculus (2nd Ed.) – Practice Problems, Methods, and Solutions, Springer Nature.

Index

A

Acceleration, 5, 11, 12, 20, 30, 36, 40, 41, 44, 46, 56–58, 60, 66, 98, 109, 110, 119–123, 129, 132
Angle, 1, 2, 5, 8, 11, 12, 17–22, 46, 74, 100, 127
Angular velocity, 93, 102
Apex, 17, 20, 21, 31, 33
Average velocity, 13–16, 25, 26

C

Center of mass, 76–79, 87–90, 102
Centripetal force, 47, 65–66
Collision, 73–79, 81–91
Common plain, 4, 5, 9
Conservation of mechanical energy principle, 50, 70–71, 76, 86, 100–102, 113–116
Conservative force, 49–50, 69
Cross product, 3–5, 9–10

D

Dot product, 1–3, 8, 10
Dynamics, 93–103, 105–118

E

Elevator, 46
Equation of motion, 5, 50, 94, 119–123
Equilibrium state, 45, 53, 122

F

Free fall, 17, 30
Frequency, 51, 71, 121, 124–127, 135–137
Frictionless, 39, 41, 43, 44, 46, 74, 76

G

Gravitational potential energy, 81, 82

H

Heuristic approaches, 5, 10–11
Hooke's law, 51–52, 71–72

I

Inclined surface, 43, 44, 46, 100–102, 113
Inner product, 1–3, 8, 10

K

Kinetic friction constant, 43
Kinetic friction force, 70, 103, 117
Kinetic, potential, and mechanical energy of a simple oscillator, 123–124, 133–134

L

Linear density, 77, 78
Linear dynamics, 39–72
Linear kinematics, 13–23, 25–37, 63
Linear velocity, 66, 93, 100, 105, 111–115, 118

M

Magnitude, 2, 8, 44, 75, 120–123, 128
Maximum range, 19
Mechanical energy, 21, 70, 81, 86, 123–124, 133–134
Moment of inertia, 94–98, 100, 102, 103, 106–114, 118
Momentum, 48, 66–67, 75, 76, 83–86, 101, 114

N

Newton's laws, 39–43, 53–60
Newton's laws in an elevator, 46, 64–65
Newton's laws on inclined surface, 43–46, 60–63
Newton's second law, 44, 45, 53, 54, 56–65, 110
Normal vector, 9, 10

P

Parallel axis theorem, 97–98, 108–109
Peak point, 17–22, 31–34
Pendulum, 74, 75, 86, 125, 135
Perfectly inelastic, 75, 76, 84–86
Period, 121, 124–126, 135–137
Period and frequency of a simple oscillator, 124–126, 135–137
Perpendicular, 3–5, 8, 9, 101, 103, 111, 114, 118
Position-dependent force, 49, 67–68
Position equation, 23, 94
Position-time curve, 119, 120
Potential energy, 49, 67–68, 81, 82, 123, 134
Power, 23, 36–37, 102
Projectile, 17–22, 31, 32, 35
Projectile motion, 17–22, 30–36

R

Range, 17–19, 31, 32
Relative motion, 16–17, 29

Rotational dynamics, 98, 109–110
Rotational kinematics, 93–103, 105–118
Rotational kinetic energy, 99–100, 103, 111–113, 117, 118

S
Simple harmonic motion, 119–126
Simple oscillator, 121–125, 127–137
Spring force, 51–52, 71–72
Static friction constant, 42, 43
Static friction force, 65
Stationary state, 54, 61, 74, 98, 102
Stiffness constant, 51, 52, 71, 73, 122, 132, 135, 136
String, 39–41, 46, 54, 56, 57, 64, 109
Surface density, 78, 79

T
Tension force, 39–41, 44, 46, 54, 64, 109
Theorem of Pappus, 76–79, 87–91
Torque, 94, 106
Trajectory equation, 22, 35

U
Uniformly accelerated motion, 17, 30

V
Vector, 1–5, 8–10, 69
Vector product, 3–5, 9–10
Vectors and coordinate systems, 1–5, 7–12
Velocity, 5, 11, 12, 15–17, 19–21, 33, 34, 36, 48, 50, 70, 74, 75, 86, 100, 101, 103, 120–123, 128–130, 132
Velocity, acceleration, and force of a simple oscillator, 119–123, 127–133
Velocity-time curve, 13, 14, 25, 26

W
Wall of death, 47
Weightless, 46, 94, 97, 98, 109
Work-energy theorem, 50, 66, 69–70, 102–103, 116–118

The manufacturer's authorised representative in the EU is Springer
Nature Customer Service Centre GmbH, Europaplatz 3, 69115 Heidelberg,
Germany. If you have any concerns regarding our products, please
contact ProductSafety@springernature.com

Printed and bound by CPI Group (UK) Ltd, Croydon, CR0 4YY

23/10/2025
01983322-0001